PSYCHO

THE FORCE OF HEREDITY

BY

CHRIS GRISCOM

Published by
Light Institute Press
Galisteo, New Mexico USA

PSYCHOGENETICS
THE FORCE OF HEREDITY

05 04 03 02 01 00   0 9 8 7 6 5 4 3 2 1

Published 2000

Library of Congress Card Number: 00-104720
ISBN 0-9623696-7-5

Cover Photography by Crash Captain, A.K.A. Bapu Griscom
Insert Photograph by Allison Ragle
Cover Design by Edward Buckley
Design & Typography by Alex Cassidy
Illustration 1 by Anja Nitsche
Illustration 3 by Edward Buckley

Published by
Light Institute Press
HC75, Box 50
Galisteo, New Mexico 87540 USA

DEDICATION

I dedicate this book to my global Soul Family who have chosen to be present at this auspicious moment in this place we call home.

Now is the time to remember each other and ourselves.

OTHER WORKS BY CHRIS GRISCOM ...

BOOKS

Soul Bodies
The Ageless Body
Feminine Fusion
The Healing of Emotion
Time Is an Illusion
Ecstasy Is A New Frequency
Ocean Born: Birth as Initiation
Nizhoni: The Higher Self in Education
Quickenings: Meditations for the Millennium

VIDEO TAPES

Knowings
The Ageless Body
Death & Samadhi
Windows to the Sky I: Light Institute Exercises
Windows to the Sky II: Connecting With Invisible Worlds

AUDIO TAPES

The Dance of Relationship/La Danza de las Relaciones
Ecstasy Is A New Frequency
Sense of Abundance
The Creative Self
Parent/Child Relationships
Transcending Adversity
Death & Samadhi
Desert Trilogy
Knowings

TABLE OF CONTENTS

ACKNOWLEDGMENTS

I joyously thank my Higher Self for transmitting to me the concept of psychogenetics and illuminating the intricacies of true heredity.

I am profoundly grateful......

To my own family constellation for making me ponder the deepest questions of relationship and heredity.

To my father, Leonard Johnston, who showed me the beauty of nature.

To all the clients whose revelations of heredity helped me see the hologram.

To the facilitators of The Light Institute for their loving, magnificent work and to The Light Institute itself for giving the world its gift.

To Navjit Kandola and Nizhoni for joining me and bringing me the joy of Soul family.

To Alex Cassidy for so brilliantly formatting this book.

To Edward Buckley for creating the beautiful cover.

To Bapu Griscom for being dragged out of bed so many times to catch the light and capture the cover photo.

My deepest gratitude to Allison Ragle who has been the inspirational and instigational energy of this book. Her heartfelt desire to share this new information with the world is the force that has brought it forward. Thank you, thank you Allison!

Chapter 1

PSYCHOGENETICS: THE BEGINNING

All the world has awaited this new, third millennium...

Looking backward, we flinch at what we have created, at what our children will inherit from us. Within our beings, we long to begin anew. Amidst the crises and chaos of today, we pray to lay down the burdens and errors of the last thousand years. Our wars and fears, our separation, our hunger to feel connected, leave us in a torment of confusion. The truths of our forefathers are lost within the confines of our scientific conclusions and we cast about in the seas of conflicting probabilities for something to call our own.

Caught in the turmoil, we ask ourselves, "Who am I?" Searching the faces of our parents and siblings, cousins and friends, we reach out for something familiar, something we want to emulate or that reassures us that we belong. Our family serves as a virtual constellation of physical, emotional and spiritual energies that surround and encompass us with the precise predilections that encourage the direction of our lives, yet we remain unconvinced and uncommitted as to the purpose of our existence.

As never before, we are left unresolved and unsatisfied within the constellation of our family in terms of the answers to our identity. The emotional struggle of getting what we want or think is our due does not come easily to us. The price of our unresolved family issues is the "ripple out" effect whereby themes repeat themselves within any other relationship situation we create, whether it is personal or job-related.

For some of us, the members of our family are our deepest adversaries or the most disengaged of strangers. Unlike strangers, we cannot seem to truly leave them behind. We attempt to extricate ourselves from their presence, but the imprints left in our psyche from familial relationships haunt the reflections of all others and hold sway over our emotional perceptions of our selves. In defiance, we rush out into the void and gather to us anyone who is also searching and is willing to become our accomplice in constructing a new identity.

Every human being is inevitably born into this process of self-exploration, which teaches us how to think and feel about ourselves in relationship to our place in a family, society and world. Up until now, we have not understood the "force of heredity" as it molds our lives. We have been oblivious to its grip, except when we hear about "hereditary" diseases or when someone points out to us who we look like. We cannot imagine the ramifications of "look-alikes" or what it means to us if our aunt has a bad experience. Most probably, we rarely ever think about extended family except at holiday get-togethers, or until some tragedy occurs.

Yet, we are inheriting minute by minute the crystallizations of our family constellations. Within our families, we are synergistically picking up emotional precepts and spiritual energies through our psychic antennae and attaching them to our DNA strands, causing us to inherit and pass them on in an endless repetitive pulse.

We can change this scenario. We can even change our genetic makeup. We can disinherit negative traits and replace them with magnificent evolutionary DNA through the power of *Psychogenetics!*

Psychogenetics is a pathway of self-exploration that allows us to examine the most minute physical, emotional or spiritual attribute that affects the overall design of our beingness. Through it, we can become aware of our inheritance as a starting point upon which we have built our sense of self and our reality.

Our sense of self begins with the obvious facts of inheritance—our physical features as our genetic relatives have passed them down to us. We measure ourselves against our sisters and brothers: who is the strongest, the prettiest?

Who do we look like? We often identify most with the ones we resemble. It is fascinating to contemplate whether we become more like them because we look like them, or we begin to look more like them because we are inherently so similar.

Look around at your family members and see whom you most look like and how you identify with them. You will note that, in actuality, you are a composite of several members of your family and extended family. You may have your grandmother's nose, your father's mouth and your aunt's torso. You may have inherited more from the members on one side of your family in terms of physical attributes, but more emotional or mental qualities from the other side.

We are more than just physical bodies. We know that we take on attitudes and styles of behavior as well as belief systems from those around us; we simply have not had a way to describe these more subtle attributes in terms of inheritance.

Familial and cultural environments serve as laboratories in which we encode learned patterns into our genetic material. This environmental inheritance infiltrates the genes and chromosomes so that we become replicates of those who have gone before us. The purpose of this genetic engineering is based within the laws of karma and can be modified through consciousness.

Psychogenetics is the revelation of how the human psyche organizes itself into a specific matrix of emotional and spiritual predilections that are as mandatory to any individual as are the physical ones. It is a breathtaking new exploration into the essence of who we are, and it will illuminate the deepest imprints stored in all human beings.

It is a story of observation. It holds within it the deliciousness of that kind of surprise which focuses our attention on something utterly new and at the same time causes us to comment to ourselves, "Yes, I knew that. I've seen that!" When we become aware of the labyrinth of energetic channels that form our unique hologram, we realize that our elusive intuition is intact and that what we sensed of ourselves is not only true, but that we can actually trace it back to its source. In doing so, we become free from the bondage of unconscious inheritance.

Our story of heredity illuminates the passage of energies from the body of humanity into the ethers where they are reassembled to suit the Soul's purpose and sent back through the veils of the unmanifest to surface again and again through the mechanism of familial genetics.

The most illuminating truth is that we are more than genetic relatives in terms of physicality. We are "Soul families" who share emotional and spiritual attributes that we have garnered through countless incarnations of the Soul. These more subtle encodings are bound to the physical DNA helix at the threshold of matter.

Whether we view this process as direct inheritance along a linear chain of ancestral seeding, or as a holographic reception of hereditary factors acquired from our own many incarnational bodies does not diminish the mystery of our genetic blueprints.

Psychogenetics provides a matrix of genetic encoding that directs us to the reference points of our interacting physical, emotional and spiritual inheritance. Once we are able to perceive the complexity of our DNA genomes, we will be able to consciously select specific genetic material that most benefits us.

What will it mean to our future to have the capacity to change the genetic bonds that limit us? I am not speaking of some scientific restructuring by means of gene manipulation, perpetrated on us from the outside; I am referring to *conscious genetics*, whereby we ourselves take responsibility for who we are!

The power of our choices will enable us to actually hone our bodies into perfect vehicles of cosmic proportion. We will use the technology of consciousness to communicate perfectly our ideal and instruct our DNA through the frequencies of light to implement the design.

When we are able to perceive the web of inheritance, for the first time ever we humans will be able to truly extricate ourselves from the black hole of mere existence and step forward into the brilliant light of freedom and choice. We will "re-invent" ourselves in a form that makes being human a rapturous adventure.

The Primordial Body and themes of power exemplified the first millennium of recorded history. The second produced emotional reaction and the emergence of the Emotional Body. The third will be the millennium of the heart. Through the integration of the holographic mind and the wisdom of the heart, we will experience a new kind of life force that will bring ease and grace to the incarnation of the Soul. A new era of cosmic consciousness will dissolve our present reality through awareness of cosmic laws.

We are entering into this next millennium with a new concept of humanity and its place in the cosmos. Not only are we clearing away the destructive genetic refuse that impedes our evolution, but we are now accelerating the emergence of a new human species with a genetic blueprint that will allow stellar travel and communication through a human body more replete with light frequencies and cosmic consciousness.

We have entered a collective incubation of suspended animation in which we dream enlightenment and the powers of creation. Soon after the commencement of this new millennium, we will realize that embodiment is for the purpose of the evolution of the Soul and a flood of new spiritual energies will engulf human existence.

The New World will not come to us. We will have to call it forth through the power of our commitment to global consciousness. Until recently, we have experienced our lives as singular and isolated realities. Now we are in the midst of the initiation of oneness. One of the most poignant lessons of the last millennium has been the illusion that we are separate beings and countries.

We are just awakening to the realization that what happens in one part of our beautiful planet affects all other parts of our planet. Therefore, we have begun to look with more discerning vision upon the attitudes and the activities of our Earth family in different parts of the globe. This mutual interdependence is part of the primary prerequisites of a safe and peaceful world where humans work together for the good of the whole.

We are not destined to inherit the weaknesses, failures and diseases of our ancestors; we are destined to help each generation enhance our gene pool through conscious scrutiny of the result of our own hereditary ingredients. *Psychogenetics* will revolutionize embodiment—forever!

This is a beginning for each of us and for every person on our planet. It is a very auspicious moment because we humans are stirring after a long holding pattern of "automatic pilot." We have been bound to our Soul friends to whom we've been born, to whom we are related—through our blood, through our cultures and, most profoundly, through our choice—the choice of our Soul.

We have blamed them and we have accused them. We have played small with them—we have *inherited* them. What has happened to our great grandparents has happened to our grandparents, has happened to our parents and has been passed to us. For the first time, we are going to have the opportunity to stop the cycle, the spin of karma, and look at what it is that we have inherited—what it means to us. For the first time, we will begin to consciously choose the inheritance that benefits our highest good.

My Higher Self has said, "What we experience will be passed down." The Native American says, "You are responsible for seven generations into the future." Responsibility is a part of inheritance, a part of the passing on. It is aligned to the imprint and the result of our choices. What has happened in our lives is passed down to future generations of humans, whether we have children or not. The echo of inheritance reverberates through the genetic matrix of our entire species. It is time to begin sorting through our vast historical repertoire and selecting that which is truly worthy of future reiteration.

From a spiritual perspective, the pulse of evolution will allow us to hand back to our ancestors the best of what was given to us as we find them again and again, lifetime after lifetime. The circle completes itself and spins up into the spiral. Perhaps for the first time as humans, we can make choices that free us from the effect of negative experiences—not only our own, but those of our forefathers and those of many incarnations of the Soul that distract us from our true destiny.

Over the last few years, I have experienced the whisperings of my Higher Self who has been showing me the matrix of inheritance. It is astounding to see exactly how and what we have inherited from our blood relatives and (every bit as fascinating) from relatives who have joined our families—with whom we do not share blood, but with whom we share destiny—who have contributed to us.

As you look at your family constellations, you will see the hologram of inheritance as what has been given to you from above and below, past and future, pulsing and moving together in one cohesive cosmic breath. You are a point of reference to all humans and that point is not just a concept of inheritance; it is a flashing access point of entry and exit!

The matrix of your inheritance is composed of revolving angles that commence pathways of relationship through which you bond likeness to brothers and sisters, uncles and aunts. From each, you draw a multifaceted composition of physical, emotional and spiritual attributes that further your divine intention.

You prearrange this destiny so that it is not accidental, but is aligned to something very profound, which causes you to think the way you do, to have the experiences you have, to be willing to move along specific confluent lines.

The thought forms you inherit from your family or your culture are the defining latticework of probabilities and possibilities in your life. Your blood and Chi inheritance, cultural and global inheritance, as well as the different qualities and specific aspects they express, are energetic predilections grouped within coordinates of physical, emotional and spiritual DNA. Once you are able to isolate these multidimensional reference points and thought forms, you will be ready to enter into the most monumental task of your life: choosing the energies that serve your evolutionary destiny and which reflect your true being!

Chapter 2

ANCESTRAL CHI
THE MISTS OF THE PAST

You are the very beginning of a breathtaking dawn, a whole new chain of human beings who will borrow bits and pieces of who you are and arrange them in an infinite array of compositions that form their source. All that you experience will become your legacy—a pathway into the future, upon which will walk a thousand relatives not yet born. You are already an ancestor awaiting your descendents to re-spark the flicker of your passage. Just as you have found your way through the footsteps of those who came before you, they will search the meaning of the prints you have left them, to discover themselves.

Imagine wandering down the paths of your ancestors. With each turn, you would discover something they did or the way they were. Flickers of their experiences would move before your eyes and each new illumination would reflect an untold facet of yourself. It might be an attribute of strength, a patterning of gestures, or even the shape of your body. As you pierced through the mists of the past, you would be astounded by the subtle or perhaps direct resemblance you hold to some distant relatives. You are bound to them through the physical heritage of DNA structure, as well as the much more elusive helix of emotional and spiritual DNA.

The Chinese describe these hereditary attributes as "Ancestral Chi." Chi is the infinite energy that sources life and it is given through the embodiment of the ancestors, who are revered as the fountain that nourishes our very existence. Ancestral Chi depicts inheritance as not only our physical attributes, but also who we are in every aspect of self. It includes characteristics that reference personality, behavioral tendencies, and even probable choices. It describes a person from a familial or group perspective, rather than as a separate individual. Our mental, emotional and spiritual attributes are viewed as part of this inherited Ancestral Chi.

Ancestral Chi is inclusive of all the ancestors. It recognizes that entire bloodlines—composed of maternal and paternal grandparents, great-aunts and uncles, cousins and second cousins—are inextricably interwoven into our genetic fabric. Relatedness is holographic so that in the great circle of family, we are touched and even branded by each and every other person who carries the blood of our relatives.

To some extent, a majority of our Earth's cultures view familial inheritance from this perspective of Ancestral Chi. They feel that the fate of our ancestors very much influences our own fate. Many cultures subscribe to the mandate that we are born into a level of society delineated by our family's position and that is our destiny or *karma*. Because of their influence, we will inevitably reflect their lives and our duty is to not dishonor them through any variation on the theme. We are given all the parts of the puzzle at birth and it is for us to accept with dignity that which is ours to live. They feel that we must honor our ancestors by living our lives within their footsteps.

The powerful influence of Ancestral Chi whispers greatness or weakness down the pathways of our possible realities. We inherit strengths and weaknesses that bind us to our ancestors and shape our lives. The Ancestral Chi communicates in perfect message codes the information that links past to future generations. These ancient connections flow from the DNA blueprints into familial channels that may be horizontal (as in siblings), vertical (as in parent to child), and even diagonal (as in aunts, uncles, nieces or nephews).

Hundreds of years ago, the Chinese devised an acupuncture treatment for expectant mothers that could counteract negative Ancestral Chi, as well as points to build the Chi. In the third and sixth months of pregnancy, certain kidney, conception vessel, and other acupuncture points were stimulated to give the child the strongest Chi possible. There are also powerful points that strengthen the Ancestral Chi, which are specifically for men. The kidney Chi, or energy, is considered the source of Ancestral Chi because the kidneys are the deepest organs of the body that pass the life force to the heart. The kidneys are also depositories of fear, and we shall see through the unfoldment of this discussion of *Psychogenetics* exactly why and how we have inherited such profound fear from our ancestors.

We can begin by contemplating their lives. Consider the realities of our ancestors. Pursued by antagonistic neighbors, unpredictable forces of nature, and extremely short lifespans, their legacies were deeply etched with fear and shadow vision. *Psychogenetics* will show us how their experiences imprinted themselves into the genes and chromosomes that produced the matrix of our future design. Unaware that our innate fears were spawned by the very real dangers in their lives, we unconsciously re-enact their conclusions of a world in which fear seems an intelligent ally.

Many cultures believe that the Souls of our ancestors are still somehow around us. They inhabit certain places or even levels of existence in worlds of after-lives. From those places, they continue to impinge upon our circumstances. Some people feel as if their ancestors are listening to their every word and are waiting to punish any misdeed. The difficulty in this concept is that it presumes we will stray because we are inherently unworthy to the task of being good. If we are expected to fail our ancestors, we probably will in some way. The future of our societies depends on our internalizing goodness and responsibility, not the impotence of external control that could shame us into compliance. To wallow in the fear of imagined improprieties is a negative reflection upon our ancestors who, in fact, might have showered us with love.

It is true that spirits may linger in the lower levels of the astral dimension, which is in simultaneous space with the third dimension. This occurs when the spirit passed out of body through some shock or violence, as well as when, for karmic reasons, it is held there because of attachments to those still in body. The cause of this may be in either direction—the survivors cannot let go or the departed cannot. We don't realize that we will be able to work out our karma with them better if we truly let them go. When we hold onto them because of our anger or our love, we imprison ourselves and constrict our own evolution.

In truth, a spirit who, for whatever reason, is entrapped in the astral has no awareness of anyone around it. Though we couch our attachment to those who have died in tones of reverence for the peace of the dead, it is more enlightened to learn how to release them into the light.

If we circumvent time and include the perspective of "incarnations of the Soul," we can imagine ourselves incarnating again and again with those we have known in other lifetimes. Each incarnation creates new karma through our "unfinished business" together. Those Soul families who have been our ancestors may become our children, as they are magnetized to us through our ongoing relationships in many lifetimes. Thus, we spin the wheel of life as we honor them as ancestors; they are pre-disposed, by cultural and familial mandates, to honor or be bound to us in their future lives. The genetic encoding passes down, well imbued with nature's laws of repetition, though dangerously ensnared in the limitations of history repeating itself.

Soul families, choosing again and again to reincarnate together, can create inbreeding that is just as devastating as that from blood relatives. The repetition of Soul themes, played out with the various participants exchanging roles such as victim and victimizer, strangle the double helix into perverted and convoluted pairings, with the predictable result of imbalanced offspring just as surely as occurs in stifled gene pools.

We never seem to finish our conversations about who we are and our interpretation of how to be in body. Like the tiger chasing its tail, we spin around and around on a horizontal plane, looking for the answers from where we've been, rather than from our infinite source. Our forefathers have given us an infrastructure that is a safety net to protect all that has been accomplished. Now, we must make the catapultic leap into new realms that will enrich our embodiment and call upon us a bright future.

OUR FOREFATHERS

In the west, we would correlate Ancestral Chi to a conversation about our "Forefathers." When we discuss them, we are more prone to viewing what they did as a description of who they were. In an odd twist, our sense of them through the "doing" allows much more freedom and possible change. The Eastern cultures view their ancestors from the perspective of "being." They therefore have much less changeability and see life from a much more fixed position. In their world, the "Wheel of Karma" seems to hold one in the place to which he is born and thus the burden of family weighs much more heavily on both past and future generations.

All of us are curious about our forefathers and when we hear about some attribute of theirs, we immediately check to see if we have inherited it. We can sense inside ourselves if it belongs to us.

My father told me that we were descended from the bloody Black Douglas Clan in Scotland that forced the King of England to sign the first Magna Carta. His description made them appear to be a wild and violent group with not much in their favor besides brute strength. I wonder if my stubbornness is an attribute of that same determination that led to their success and I hold the expectation that our rather rugged constitutions will bear out the positive adaptation of their physical strength.

Shakespeare said that "The evil men do lives after them; the good is oft interred with their bones." On the emotional front, we are quick to point out how certain family members have inherited a "bad temper" from their rather infamous distant relatives. On the one hand, it is a foolproof justification of horrendous acts, and on the other, it provides the perfect weapon for mudslinging between family factions. Through consciousness, we can learn to use inheritance as a guidepost that can show us, in advance, how to embrace or alter our predilections.

Anything that happened to your grandparents or your great, great grandparents is a part of your family inheritance. In some family members, those attributes will come forth in a very direct manner; in others, certain aspects will be completely hidden or recessive. Their stories make for fascinating contemplation when placed in the realm of reference to ourselves.

I remember when we first did Family DNA sessions at The Light Institute. I was asked to access a spiritual inheritance within my family constellation. Up came my great, great grandmother, whom I know absolutely nothing about. She was magnificently real! I saw her up close, as she appeared right in front of my face. I could feel her breath on me and I clearly heard the tone of her voice. The energy she carried as a result of her experiences felt as if it were my own.

When I went back into the lifetime that held the source of those energies, I could look through her eyes and heart and I knew her from the deepest level of her being. I was so thankful to have her in my family tree, but also to release her burdens from my own soul patterns, for they were ensnared in the consciousness of those times.

It is just the same when we see ourselves as a killer in sessions and we go back into the childhood to find out what made us that way. As we look through the heart and emotions of that person we have been, we can completely comprehend why it was that way for us. We can release the judgment we held on ourselves because suddenly we feel how we had perceived the world. We did not see a way out and therefore we had thought we were justified. We did what we did because that was the choice we had at that time.

Similarly with our ancestors, as we go back into their lives, we can see them from a different perspective—as they were, as they lived. Because we have accessed that history *energetically*, we are able to dissolve it from our emotional and spiritual DNA.

Looking through my spiritual DNA, I found a beautiful inheritance that has been a source of the work I do today. I saw my great grandfather on my mother's side of the family. He lived with us when he was ninety-two years old. He is the son of the great, great grandmother whom I saw so vividly. I have always felt him to be one of the most important people in my life, partly because he loved me so much and partly because he was so brilliant and alive. His charisma filled all the space wherever he went. He was very thin and had long white hair and a long white beard. He had to gum his food because he was missing so many teeth. Perhaps because of his Indian heritage, he was a great storyteller and when he spoke, I was immediately transposed into the scenes of his tale. He could carry me equally as well into a Shakespearean play while he perfectly quoted all the parts, or into the horrors of Wounded Knee, which he related from the perspective of the Indians.

He carried a cane and used it for emphasis when he told his stories. I remember him telling me a story about the stars and he emphatically stomped his cane on the ground.

"You must bring the sky down here, little chick!" he would say, pointing his long bony finger at me. I shuddered as he pierced my being with his stare. Time stood still. I knew that he was saying something important to me, but I didn't know what it meant.

I grew up to find the "Windows to the Sky" of Chinese esoteric medicine and to learn how to "Anchor the Sky," which has to do with activating our Galactic DNA through the spin points in our body. I know now that he was "seeding" me from the deepest reaches of his being and that it is something we share on a Soul level.

When we are doing sessions, a beautiful memory packet may surface and you may want to hold onto it. Initially, it is difficult to comprehend how letting go of a powerful experience will bring the energies back to you in a free form so that they can affect your life now.

The memories act as energetic links to the experience or lifetime you had. There are other memories or levels of consciousness that are also stored in that lifetime, which you would not want in your present existence. Some deeds, for example, were perfectly acceptable then, but would be seen as barbaric and inexcusable now.

Thus, it is essential to dissolve all the points of reference to incarnations—their content, their energetic residues within your blood crystals and DNA. In so doing, your consciousness can become part of the creative force—allowing you to continuously reinvent and recreate yourself through the infusion of talents your Soul has accrued in its sojourns into body.

We do not have to carry the past as a burden of inheritance; we can carry the past only as a point of illumination that synergistically instructs us within the laws of cause and effect—karma!

NAMES

The glue of inheritance is not just the physical or emotional imprints from ancestors. There are many elements of mindset and lifestyle that shape the bonds of family. Even the Surname, or names given to honor family members, influences our sense of belonging.

Consider the power of a name. What could it mean to you to truly fathom that there are people around the world who share your family name? You may have never contemplated how your name creates a link with all those people. Through that sound which identifies you all, you are energetically connected. Wherever your name originated, it was carried out around the world, jumping bloodlines, cultures and races. There is a certain commonality between all the people who carry that name.

My mother's maiden name is Smith. On one of my favorite islands in the Bahamas, there are whole extended families with the last name of Smith. I wonder how that name came to these small islands. Perhaps some slave trader, hundreds of years ago, named his captives "Smith." Neither my Bahamian friends nor I would claim any relation to him at all, but history, as well as blood, weaves the human fabric of relation. Because of the name, he is an aspect of me, though I know nothing about such a person. He is an aspect of them too, though their only connection may be an arbitrary bestowal of a name. That point of history has instilled a coded memory that links us together.

When I first met the Bahamian Smiths, I liked them right away and I felt such a delight in telling them that we shared a name. I smiled as I watched them look me over in an almost imperceptible way while they turned over in their minds who I could be to them. They are from a different culture and a different race, but we have an irrevocable link somewhere in our distant past that brings us to the energy of family.

It is interesting to consider the genesis of our

names. In Germany and most of Europe, last names originated through some description of the place or the occupation of the family ancestors. For example, Bauer means farmer, Weber is weaver, and Koch is "cook" in German. Thus, the name holds the energy of specific attributes that described a family and its designated function in a village or tribe. As we carry the name, the skills could be points of reference for us now. They may be seeds of our repertoire that we can utilize today in more subtle ways. Imagine the orchestration of pattern and form involved in weaving a textile. That same skill could resurface as the art of a successful CEO or mathematician who recognizes how all the small pieces create the whole.

Names hold certain frequencies of sound that become the calling card of your being. Spiritually, each person picks their name, just as they pick their parents. It is very important to align to the sound of your name. Whether it is your first or last name, it is the sound of the self. You have whispered it to your parents and it resonates to a vibration, which fits your evolutionary pattern.

Your name could be one that you carried in another lifetime, or that you associate with someone you have known. Everything that has ever existed or been manifest in the universe creates an imprint. We all relate to those imprints without consciously knowing why. Parents often name their children after people they admire, expressing a hope that the child will also be like that person. The child, on the other hand, may hold other associations with the vibration. The name may be of an enemy in another lifetime and their painful resistance to it may be exactly that which attracts the parents to choose it. My Higher Self says, "Whatever you resist, that's what you get!"

How do you feel about your name? If you like it, you have probably accepted yourself. If not, you are perhaps still seeking some way to feel more powerful and desirable to others. You may despise a name that reminds you of a person you don't like, or be distrustful of another person with that name, simply through association.

Names are like clues that help you unveil the mysteries of inheritance. You are linked to all those people who have carried your name. Yet, you do not need to resist or labor under the burden of their lives. They have their own stories and only those themes that you share, will converge into the rigidity of inheritance.

Young people often shorten each other's names to add an element of special intimacy and a sense of independence from their parents. When someone says that they "hate" their name, you can be sure that they feel separated from their family and themselves. That name vibration must be a reference point to some experience that is locked in their genetic memory banks and is triggered by the sound of their name.

You may feel connected to a person who has the same name as you. In reality, you have inherited and are an open channel to everyone who carries or has carried your name, even those who are not connected to you through blood. Consider the certainty that around the world, there may be millions of people who have your same name!

In many cultures, women are expected to give up their last name and join their husband's family when they marry. Today, women are choosing to keep their own surname or are creating hyphenated combinations that represent both families. It might be interesting to check the numerological system of letter values to see which combinations give the most balance to your name vibration. When I became a Griscom, by marriage, my sound frequency was strengthened and balanced.

You can come to embrace your name with all its inner correlations. Furthermore, through your life choices you can actually give back energy patterns to that vibration which will echo down the pathways of inheritance to untold future generations. You will impart to them the illumination of your consciousness and the joy of life!

Chapter 3

THE DOUBLE HELIX

We are so beautiful! Our essence is a design of chemical pairings anchoring a twisting, swirling ladder that rises up in a perfect double helix pattern. This exquisite fluid essence of life is a protein molecule called DNA, deoxyribonucleic acid. It is the chemically-based structure of the genes and chromosomes that sculpts all living form. We humans have an encoding of 50,000 genes, which combine themselves in a fantastic array of variables that provide for our uniqueness. No two humans are alike.

Until recently, we have been given the impression that genes and chromosomes are set and inheritance is a fixed arrangement that cannot be changed except by possible genetic manipulation. To the contrary, there is so much impinging on our genetic encoding that alteration is absolutely inevitable; our essence material, our DNA, is in an eternal flux of evolution.

Cosmic pulses that have altered life for eons are quickening Earth's evolution. Some would view this as only a phenomenon of natural forces, but I am convinced that it is a conscious act of higher beings who have purposefully interjected specific material into our DNA to lift our species onto higher planes of existence.

We have had many "seedings" that have accelerated and infused our DNA with energies that bring us into awareness of all that is around us. The loving energies of Christ, Buddha and other divine masters are the supporting structures of our spiritual DNA and have come in waves of new experiences that encompass humanity. Spiritual DNA is infused with the strands of our emotional and physical DNA and provides the mechanism by which mutation becomes purposeful intention.

There is a universal consciousness that guides our evolutionary process. It has brought us to a cyclical point that we come to every few thousands of years, in which new energies reshape what it is to be human. We can quicken that process right now by learning how to access and clear our DNA material. New genetic influx always causes a flushing of the old unworkable vestiges of past constructs that run counter to adaptive refinement.

With the advent of genetic exploration, we are discovering how the genes are susceptible to environmental influences that alter their makeup. While we sadly face the results of our environmental pollution, it is destroying whole species and the evidence is mounting that we are not excluded from its devastating clutches. Not only our environmental meddling, but also our experimentation with mind-altering drugs is plaguing us with the inability to conceive, deformed fetuses, and genetic diseases caused by the disruption of helical flow.

There is a way out. We can learn how to communicate holographically with the consciousness of life. What we see under microscopes will never show us purpose or the intrinsic links between matter and its essence. They are not held in visual categories, but rather, they exist within the finite relativity of conscious energy. The solutions can only come through the whisper of cosmic guidance. If we will allow ourselves to reach beyond the constriction of technology and form, we will find the answers as to the "why and how" and we will re-invent ourselves in accordance to divine laws.

The concept of physical DNA is something that we have known about all our lives. We think we understand inheritance from the point of view of chromosomes and genes, but we have not learned their elusive secrets. Our bodies remember. They remember the bodies of our mother's mother's mother. They reiterate the experiences and outcome of those forms so profoundly that we cannot truly tell the difference between our own and theirs.

Physical heredity is the creative expression of something beyond form, something that holds knowing and feeling of self. In truth, the separation between physical, emotional and spiritual energetics is only an arbitrary delineation that does not actually exist. They flow into each other in ways we may never capture under a microscope because of their dynamic interplay.

Only by acknowledging their relationship will we discover the answers to mysterious genetic mutations. The variance between them is the same descriptive shading as between biochemical and electromagnetic currents. They may seem quite different at first glance, yet both are central to life itself; they are intertwined. Physicality threads back through electric-fluidity into light, which sources life.

Physical DNA is inextricably interwoven with the subtle fabric of emotional DNA. The emotional DNA piggybacks the physical and plays sentry to bodily will, influencing it through long-standing emotional characteristics. Though our technology is not yet refined enough to view it, we are perceiving its shadows and residues, marking the changing tides of emotional currents. Emotional DNA is equally as specific as physical DNA in terms of inheritance through familial or cultural channels. You may not be able to articulate a sense of emotional "likeness" with members of your family, but you know it is there and that it influences how you interact with everyone else.

It is actually common to hear statements such as, "You have a temper like your father," or "She withdraws like her mother." Emotional DNA can be viewed as inherited character similarities such as humor, style, passion, or typical responses to life situations. These emotional attributes are rarely seen as arbitrary, but rather as predictable patterns that you would recognize as exemplary of people in your family constellation.

Surrounding and encompassing the physical and emotional DNA is the spiritual DNA. Like an etheric encapsulation of light energy, it wraps around them as an auric field that descends into the physical through the master glands of the body. It encodes the elements of destiny and cosmic law into the pattern and potential of embodiment. Spiritual DNA is our absolute blueprint. It expresses itself in our love of nature as well as our longing for the stars. It is the genesis of mystical imprints and religious devotion.

The three DNA parallels resonate to specific energy or *chakric* centers that become the genetic spin points for them in the body. The base chakra holds the physical spin point, the solar plexus chakra holds the emotional point and the third eye holds the spiritual spin point. By virtue of its chemical nature, the DNA complex is reactive to the subtlest whisperings of conversations that occur on emotional and spiritual levels, expressing the nuances of our very Souls!

Emotional currents alter the chemical releases within the hypothalamus and endocrine system that govern our state of being and play back to the biochemistry of our genetic makeup. Here is a bridge between the spiritual and the scientific, a connecting link between structural and amorphous realities. Thought and feeling define destiny and body.

This is a profound realization! It means that we are not totally shaped by a physical world, but by a more elusive intelligence of universal proportions that sketches the conversations between body and mind. Thus, emotional and spiritual energetics may be the hidden precursors to physical experience.

Vulnerability to diseases and genetic predilections experienced by our relatives may be imported to us through emotional imprinting we have taken on from them, rather than the inheritance of an imperfect gene composition.

Inherited thought forms of fear and negativity open us up to the same diseases our progenitors experienced and may actually *cause* a genetic irregularity that then translates into physical disease. On the other hand, joyful and positive emotional characteristics would be the most powerful precursors of good health. We can discover a great deal about our genetic encoding by exploring the stories of our ancestors and those of our family constellations to see the predisposing experiences and conclusions they have handed down to us. In the future, genetic predilections will be correlated as much through emotional and spiritual imprints as by physical ones.

In my last book, *Soul Bodies*, I mapped out emotional themes corresponding to specific disease etiology. Each type of disease relates to certain bodily themes of emotional origin. The biochemistry of our bodies is responsive to environmental influences that include emotional fields as well as physical ones. The physical body is a cohesive constellation whose various organs and parts are symbolic of its functions and express themes that help us to listen to and understand the body's intention.

Since the body holds the Soul spark, its intention is always perfect resonance with the Soul itself. All disease has a spiritual purpose or reference point. In my many years of exploring traditional medicine, this has been a universal concept reiterated around the world. Traditional healers will always look for a corresponding imbalance of the spirit when they are searching the source or solution to an ailment. Disease is not considered a punishment forced on us by our physical bodies, it is witnessed as a lesson being taught by the magnificent Soul.

The "Soul Body" perspective offers us a revolutionary illumination: We are not the *victims* of our bodies; we are instead, the *designers of our destiny.* What happens in our bodies is a reflection of Soul lessons offered to us *through* the body. We can actually alter our DNA to suit our physical, emotional and spiritual purpose, once we have the consciousness to perceive it.

We can change our "genetic habits" and create new ones that help us to be more available to the energies around us now. We have never done this because we have not had the awareness of genetic patterning. Now that we can map the matrix of our blueprint, we can begin to participate, through our consciousness, in the perfection of its design.

The Light Institute sessions give us an opportunity to view inherited attributes from the practical aspect of what has happened in our lives. When we pinpoint a physical imprint that we want to dissolve from our DNA, we employ our intuitive faculties to perceive it and identify its location on the DNA strands. We ask the Higher Self to show us the DNA. It is not too difficult to "imagine" physical DNA, as most of us have seen pictures of chromosomes and DNA strands.

There is no external rule as to how the DNA should look. Intuitive sight may describe it from within its structural strands or through the fluidity of its chemical components. One person sees it as a bundle of spears; another as a twisted rope. People will perceive it in their own way. Descriptions of DNA are totally subjective and at the same time integral to the experience of body awareness. There is something very profound about seeing one's own DNA. It is a kind of cosmic messaging that informs us of our own truth.

Next, we ask the body to show us exactly where the inherited trait is on the DNA strand. One person may see it as a black spot; another may sense it as webbing or a tear. How we see it is unimportant; what is crucial is that we can identify where it is.

A laser of white light is the tool we use to clear the DNA. There are more than two hundred tones of white light. Some, such as radiation, have a magnitude of radiance that our bodies cannot withstand. Our Higher Self selects the frequency and we then laser that light directly into the point and remove the imprint.

The brain records all visions in the same way, whether you see them with your actual sight or with your inner vision. If you look at a tree or close your eyes and imagine a tree, your brain will record the tree. This capacity has tremendous potential application as we learn to apply it to our participation in invisible realities, such as our DNA.

The same process of "location and laser" is used for altering emotional or spiritual DNA. It is absolutely amazing to encounter something so entirely yours that needs no outside technological assistance to access. The sense of freedom that comes from releasing unwanted inherited traits is exhilarating and by your own experience of it, you begin to trust who you are.

How powerful to see from whom you inherited something that is a weak link in the chain and to know that you can change it. Even more inspiring is to hear the whisperings of the magnificent gifts that you have inherited, but have not yet claimed.

"Yes, I do have this capability; I can perform these things," and to recognize for the first time what you may have intuited, "Yes, that is like my father. Yes, that comes from my great-aunt." These insights free you from all bondage and carry you out to the periphery of potential!

In some circumstances you might want to *amplify* something on the DNA that enhances your well-being. Perhaps you discovered a gift of humor or healing that has come from some relative. You can simply utilize the white light to activate the quality. Stimulation of the DNA in this way often creates a physical sensation or vibration that can actually be felt in the body.

Here is an excerpt from a Family DNA session:

The inherited gift: *"It's a capacity to see beyond the seeing. Knowing what is going on inside someone; how they are feeling and a strong sense of intuition."*

"It's in the spiritual DNA which looks like a glistening, twisting umbilical cord of translucent bluish-white. It's a kind of signature, like a webbing of opaque white within the membrane."

"Lasering it is causing an electric current to flow through me. I feel electrified. It makes me ecstatic!"

From the perspective of a fluid, changeable double helix, inheritance takes on a new meaning. You can learn with a new openness and respect about your ancestors and the indelibility of their experiences. The tapestry of familial contribution weaves a tale of exquisite beauty that, in effect, has no beginning and no end. You realize that you are truly cosmic in origin and life becomes something that cannot possibly be limited to just one body, one time, one place. They live through you and you are designing the infrastructure of your descendants into the future.

When your sense of self has expanded to include the repertoire of your entire family constellation, the energies will begin to flow into each other so that you experience the "knowing" of physical, emotional and spiritual currents as a part of a whole. Simply including their experiences and thought forms as part of your source material will help break down the illusion of one lifetime and one body. Think of all the bodies you have already inherited from your ancestors!

Ultimately there is only one Soul. Through spiritual and emotional DNA, you have been a part of infinite other lives. Viewed in this way, reincarnation does not seem so foreign or implausible. This is the time on our planet to stretch all possibilities so that we can make the leap into new worlds that are hurtling themselves towards us at Mach speed. The Soul has never been a part of our notions of proof. Our DNA signature holds residues and imprints from other incarnations that must be addressed in order to catapult ourselves into this new cosmic pulse.

Whether you view these vignettes as mere stories, as lifetimes inbred in you by your ancestors, or as actual incarnations of the Soul makes no difference. Working with them as a part of self-discovery will change your life and the synergy they provide will literally open up new worlds!

Within its vast memory banks, the DNA holds all

experiences and collective history. Thus, your body intertwines the stories of your ancestors and those of the other bodies you have had into one contiguous adventure of life. It uses them all as points of reference for orientation and response. In this life, you may have suffered a profound fear and through familial or incarnational exploration, discover that the source of the fear is not in this body. It is only "deposited" in this body through the continuity of DNA.

PHYSICAL DNA

All of us are fascinated by the comparisons between ourselves and the rest of the members of our family. From the time we are children we listen to and examine for ourselves the similarities and differences between our body features. We wish fervently that we looked like someone or we hate the fact that we are so similar to someone else.

Our inherited noses, chins, bellies and legs are a part of family pride and belonging. Beauty has always been a subjective observation of refined familiarity. More specifically, our physical resemblance carries with it a verification that through genetic matching, we are somehow closest to those we look most like. The closeness comes in the form of emotional or spiritual encoding that is the result of shared history beyond this one body.

The physical body carries cues from other incarnations as well as cultural and racial heritage. The famous Roman legs that provided the strength of conquest, or the slanted foreheads of the priests of Maya and Egypt are whispers of possibilities we have never explored. To even glimpse bodies we have carried is absolutely wondrous! Sometimes it is shocking to feel through our present bodies what other bodies have experienced, but even if it is difficult, it is also amazing and great fun. Let me show you an example of how clearing a physical DNA inheritance sourced in another lifetime can have a profound and healing effect.

This is from a session of a woman in her early forties who had suspected breast cancer. She had come to our special "Soul Lessons Through The Body" workshop.

"Ask your Higher Self to show you someone in your family constellation from whom you have inherited this tendency."
"It is my aunt on my mother's side. She died of breast cancer."

The Lifetime...

"It's a war. I can feel the cold damp place where we are hiding. The smell is stinging. If I open my mouth, it rushes in like poisonous air and chokes me. It bites at my nose and hurts me so much, I try not to breathe. I am crying and crying. My mother puts a rag into my mouth to make me be quiet, but her fear feels like daggers all over my body and I cannot stop myself. I am screaming and gagging on the cloth at the same time. In desperation, she offers me her miserable, flaccid breast. There's nothing there. I cannot understand this torture. Where is my nourishment? Where are the

soft, warm forms that were mine? Now there is nothing, nothing but the sucking air that hurts my throat and these cold, flagging cushions that refuse to comfort me."

"I can cry no longer. My body is filled with a numbness that has spread through me from my mother, from life. The blurred shadows are silent now. It stops. I am gone."

"What is the thought form you have brought with you from that experience?"
"There is nothing that can nurture me.... And, there is a feeling—I hate breasts."

"Go back and look at what your mother is feeling when you die."
"She is screaming inside. She hates herself. She is furious that her breasts went dry. She blames her body for not saving me. She vows to herself that it will never happen again."

"Go into your DNA and find where that thought form of hating breasts and not being nurtured is stuck."
"I see the DNA like dried string laying as if it were unwound. There is a black spot in the middle of the coil."

"Ask your Higher Self to laser the brightest white light into that spot until it's gone."

After the session we saw that her aunt had carried that violence and anger towards her breasts into this lifetime as well and because they had shared such a painful experience, had passed the negativity to her.

She said that she had never liked her breasts. After further clearing, she felt very differently about her body. A month later, the lump had proven benign and had almost completely dissolved.

EMOTIONAL DNA

It is not difficult to imagine inherited emotional patterning as all of us copy and imitate the posturings of those around us. The most blatant stereotypical themes of victim and victimizer, saviors and healers, martyr and hero all lend excellent material to our dramatic fantasies and we can witness them being played out at any moment by the members of our family constellation. All too often they become the signatures of our self-identity.

Many pathways of DNA begin as thought forms that arise from emotional conclusions. Repeated static descriptions of who we are ink themselves into our emotional outlook, becoming as deeply ingrained as our physical features. They are not easily dislodged because they are so much a part of our expression. Likewise, the thought forms of our parents are potentialized in us and influence our emotional makeup just as does their eye or hair color.

Emotional inheritance may express itself physically in disease and even in bodily structure. Bulbous noses are cues of emotional heart energy, just as round faces illuminate emotional themes as a part of life purpose. Try matching physical features with relatives and you will find similarity of emotional characteristics.

Let's look at an emotional theme story and see how the emotional DNA is transferred.

"Ask your Higher Self to show you someone in your family constellation from whom you have inherited emotional DNA. See what it is."
"It's my Aunt Barbara. It's a high kind of manic energy, caught in a heavy depression. Yes. I have this same kind of high energy, but there is a deep sorrow that goes with it. I don't know why. I love life and most of the time I am so up that I kind of blast everyone around me."

The lifetime...

"I see her sitting in a rocking chair, just staring out in space. I'm a small little girl playing at her feet. I can feel her just as if she were talking to me. She lives with us because something is wrong with her. Everyone says she isn't there, but I know she is. She can't talk or do anything, but I love being next to her. Her body is stuck in the chair, but her mind is out in the stars. There is something so ecstatic about her, but no one sees it except me. They say she's in a stupor because of an accident when she was young."

"Somehow she is showing me pictures. They're like star maps! I'm seeing constellations and something that looks like flight patterns."

"When I get older, I feel the heaviness–the depression because it's all locked in me just as it was in her. I am afraid to let it out. If I told anybody, they would think I was just like her. I thought she was like she was because of all that stuff in her head—that the star maps

kept her frozen in her body."

"I hid it all away, deep inside myself like a dark secret. I wanted to burst out, but it was always there, holding me back. I would feel so light and free and then something would remind me that I couldn't be free because of what was locked inside me. I began to drink—a lot."

"I traveled around the globe making maps, but I knew I should make the star maps she put in my head—I just couldn't."

"I died from drinking—heavy in my body, from the weight of being bound to a world everyone else knew, but which was a prison to me."

"It's been like that in this lifetime, too. I've been a heavy drinker–trying to forget what I am afraid to let out. I tried to navigate through it by letting the stars carry me out there. I see that it isn't worth it. There must be a way to use it this lifetime. I still miss her."

"Look at your emotional DNA. How do you perceive it?"
"It looks like a clump, like a star that has fallen in on itself. There are points sticking out, but you can't tell where they belong."

"Look and see where this manic-depressive energy is caught in the DNA."
"It is as if the clump were a fabric with irregular blobs of markings on it."

"Ask your Higher Self to laser the highest frequency of white light into all the blobs to release this emotional pattern."
"Yes, it is beginning to swell up. The blobs have turned

into constellations that are connected to each other. It is like an imperceptible membrane that holds them all together. I feel very still inside, as if I've been put back together with everything working."

SPIRITUAL DNA

Unbeknownst to our limited vision, our bodies flaunt their divine source. Filled with elements of earth, air, fire, water and the ethers, they powerfully express cosmic inheritance. We have never wielded them as the cosmic tentacles that they actually are because we could not guess the intention of formless divinity. We look above, into the seamless sky and beg to be transported out of our pitiful mortality into something we dream more worthy. We have done so in all cultures and in all times. We scan our lives for the portals into other dimensions and long to be the "chosen" ones who have been granted the right to exist in divine rapture without the burden of daily life.

Virtually every culture has found a way out of the smallness that squeezes life, but none have offered it to the whole of unconscious beings. It has always been secreted away and given only to the elite. The Chinese called the spin points to expanded consciousness "windows to the sky," and the emperors and priests claimed their right of passage through divine heritage. The masses, left behind and seemingly disinherited, descended into the pit of darkness where no horizon foretold of their divine birthright.

Separation from spiritual source has proven too

deadly for us to bear and we must now find a way to revisit our origins while being present in our bodies. We must "anchor the sky" by experiencing a sense of the Divine within our bodies. All of the pure energies of love, as well as memories of enlightenment, are written in the Soul patterns of our spiritual DNA and are accessible to us within the context of our bodies; we only need to become conscious of their existence.

We are coming to a time when the galactic cosmic energies, and the sun's radiations, are moving in and through us. We can anchor the sky by recognizing these frequencies as innate to us—not something separate, but ourselves. We are the Divine. We are the cosmic rays, the source of life. We are the radiance of the stars, the carbon atoms of the sky. Anchoring the sky is simply a reference to consciousness. All these aspects continually recreate us through the awakening of our spiritual DNA.

We are suffocating beneath the constrictions of our religious dogma and belief systems that cannot encompass the whole of our true spiritual essence. We have outgrown them. Spirituality is awaiting a new human form.

In the past, we may have chosen to incarnate into a culture or religion because we ourselves had created the format of religion as the only expression of spirituality. It is not that we needed *more* spirituality, but that we did not know how to find it within ourselves.

Spiritual DNA is our divine heritage. It is as intrinsic to us as breath itself. In fact, even through our breath, we are seeding future generations with spiritual Chi that links us together psychogenetically.

Within the circle of life, we are continuously inheriting spiritual imprints from those who are our ancestors or whom we have known in other lifetimes. We may inherit traits from our children that were passed to us through their experiences and are returned now as we transmit those traits down to them.

Whether we see inheritance as something that comes strictly from the generations of our family tree, or as the result of eons of incarnations of the Soul, there is a continuous flow of energy that creates grooves in which to channel the material that shapes our spiritual destiny.

Imagine the possibility that everything humans have ever experienced on a spiritual level is somehow a part of our spiritual DNA. The religious wars, the dogma, the fear, are all part of what we must clear now in order to allow a new, more cosmic spirituality to grow in this next millennium. The unspeakable abuses that man has heaped upon man in the name of God will only come to an end when we have erased the traces of such negativity from our genetic encoding.

If we were to follow them back to their source, we would find a historic mutation in which there was an eclipse of human consciousness that lost sight of divine connection. In our fury and fear of separation, we thought to appease the "Almighty" through actions that would show us worthy. We concluded that only life itself was powerful enough to get attention. It was a conclusion that is still causing us immutable suffering today.

On the other hand, the mysticism and ecstatic experiences through which humanity has seen the face of God or felt the grace of a loving universe are likewise encoded in each of us. The saints and the mystics have also seeded us with their potential. Because they reached these heights, we are gifted access without their sacrifices. What any human has ever touched is accessible to us through our psychogenetic channels.

There are those who hold the light for us. They might be a spiritual teacher, a wisdom keeper, or a religious official of a church. Through psychogenetic channels, our grandparents pass to us the spiritual DNA that gives access to the unmanifest universal energies.

Godparents too, hold that space which ensures a format of spiritual conversation. Even the word, "Godparents," acknowledges a role inclusive of something beyond daily life. They not only promise to protect and raise our children in our absence, but to teach them about the world of the Divine and to see that they find a place there. Though we often choose our closest friends with whom we share philosophical perspectives, deep within us we have a sense that these friends are connected to the "Great Spirit."

It is as plausible to experience spiritual inheritance in a rainforest as it is within the confines of a religious order. It is not passed to us by our cultural or familial mandates alone, it resonates with our absolute essence. *Psychogenetics* gives us a tool to access that essence. Our awareness of this new biological technology can potentiate our multi-dimensional embodiment. Spiritual DNA holds frequencies that are not mitigated by human law, but by the very force of evolution itself!

Chapter 4

FAMILY DNA

There is one word in every language that has a profound effect on our sense of self. It is the word "family." Whether one is adopted or orphaned, lives with a grandparent, or has the classic triad (father, mother, child) reality, a discussion of family brings out the most intense confusion, joy, anger and longing. Virtually all of us see our place in our family from the perspective of an insecure ego. Did we get enough or give enough—love, attention, respect, things?

We cannot wait to leave home or get back home; we wish things were like they were in our childhood or we don't remember anything about it. One moment we are overwhelmed by nostalgia and the next moment, we are reminded why we don't really want to spend time with anyone in our family. We can shut out the memories and move away, but all the experiences of family are indelibly etched in our cells. We have inherited the consequences of family life as surely as we have inherited our physical DNA.

Exploring our family ties is one of the most courageous adventures we can choose because we will inevitably discover our own reflection in the attitudes and habits (as well as in the faces) of our family members. No matter how much we deny it, the good things and the bad about *them* must somehow also be about *us*. Even more profound, if we look deeply enough, we may dislodge the blame and anger we have heaped upon them as we excuse ourselves for our own ineptness and failures. What is true is that we are a part of drawing them to us, even our distant relatives. Our Soul had a purpose in choosing them to instill their genetic traits in us and when we find that purpose, we will come to love and honor them in a way we've never known before.

These familial characteristics that we so admire and detest may have been passed down from someone in our nuclear family or from a distant relative. Whatever their source, we cannot escape them and if we are brave enough, we can use them to piece together the puzzle of our undiscovered history.

If we knew what it was that we inherited from the members of our family, we would surely feel differently about them. Even if it were something negative, as we saw it clearly in both ourselves and them, we would feel a new kind of compassion that seems remote from us now, while we hold ourselves so apart from everyone else. To be like another person is an affidavit of kinship that promises a place of belonging, away from the cold winds of aloneness. It is an insatiable hunger that gnaws on the very threshold of human culture. In truth, we want to know who we are like and whoever it is, we are thankful that they are there. We are comforted by the inclusion in anyone's life that makes us more real. Each facet of our inheritance tells a story about how the invisible energies inside us create our external identity.

I have often watched my grandson make the same gestures as my daughter, his aunt. Within the family, these two resemble each other more than they do their siblings. They share many features, including facial structure, body style and emotional sensitivities. Their expressions reflect a likeness of mind. It is not that he is feminine, or she, masculine; it remains that they are cut from the same fabric, much more so than any other two members of the family. Whether they are from the same quadrant of the galaxy or the same seemingly random genetic constructs, their resemblance is fascinating to witness as he—in some mysterious way—re-enacts his aunt. I await the future with great anticipation to see how they unravel that alikeness. Perhaps they will work together or unite in some special way that augments their impact upon their family and the world.

One of the most important aspects of family and family inheritance is the way that we receive energy. From the moment the fetus attaches to the uterus, it receives its nourishment through the umbilical cord. All the feelings and emotions of the mother and the psychic energies of the father come directly into our body as it is constructing itself in the womb. Taking on the feelings and impressions directly through the blood makes it very difficult to separate what belongs to someone else and what is ours. We cannot recognize where we leave off and our parents begin.

Taking energy into us becomes a habit through which we experience ourselves. Whether we receive "enough" becomes a conversation that may distract us from maturing into givers who can sustain a true relationship. The two-year-old says, "I want that; you give it to me." Thus, we begin a process in our lives that is a dead end, that stunts our growth. In order to cast off the shackles of family "IOUs," we have to move through the conversation of what we didn't get, to what we did.

Our emotional baggage is full of what we have demanded from those in our family and the many ways we have felt cheated. I think every child says at some point, "You didn't give me what I wanted," or "You gave it to somebody else," or "I am unloved." Our reaction to those feelings is to bring forth the killer energy of punishment. Blame and punishment are two very destructive energies that poison us. Though we often feel punished by our parents, we quickly learn to punish them, as well as placing blame on them for what we do and who we are.

By the time you are four, you know exactly how to "get" your parents and siblings. You know how to defy them, deny them, rebel against them and make them miserable. You know exactly how to push all those buttons and you're still a tiny child. Because you share blood, you have the capacity to listen through the DNA into the Soul and through it, into infinite lifetimes.

As children, most of us engaged in all manner of emotional denial and resistance towards our families. When I was a child, I was sure that my parents didn't love me as much as they did my older sister. My emotional fantasies ranged from a variation of the "Cinderella" theme of the scouring cinder girl with only the animals for friends, to the high-pitched, "I'll just die; *then* you'll miss me!" Mostly it ran along the lines of "Nobody understands me. Nobody cares."

I didn't find out until much later that my sister, who seemed to be perfect, had an even more bizarre fantasy than any of mine. She secretly felt that these weren't her parents at all! Some torturous trick of fate had callously placed her here. She was the first-born and so I worshiped her in every way. She was much prettier than I, smarter and easily loved. Of course, she saw it in a completely different way.

We create whole fantasies about how our family just doesn't know who we are. These thought forms carry a stigma. They imprint themselves into our emotional makeup and we repeat them with our lovers and friends until they become a staple in our emotional diet.

For what terrible things do we blame our parents? A thousand immeasurable slights! They did not intuit our every need, did not uphold our momentary fantasy. We may be disgusted by their bodies. It might be something as simple as not having breasts. I remember at 15 being deeply angry with my mother because she had no breasts and I had no breasts. I felt I would never be a woman.

We are angry with our parents about their physical imperfections because we have inherited their form. They got it from their parents. What could they do? They have had the same dialog themselves.

Almost always, we feel justified in blaming others.

A parent might have done something to the other parent or to you that you feel justifies blaming them. From your conscious reality, it seems easy to separate the "good" guy from the "bad" guy. On the surface everything appears clear. However, until you can see the karma between two people on a spiritual level, you can never know the purpose of any experience or relationship they share—how the person who is the victim is calling forth the victimizer for their own balancing, for their own healing.

The victimizer may be their best Soul friend, whom they have called in to help them learn a lesson. Have you ever stepped between two people and experienced them both turning on you as if *you* were the bad guy? Truth is not a flat, linear representation of what we can see in the third dimension: Truth is a multi-faceted orchestration of Universal Law in motion.

Divorce is a favorite pretext for blame. Parents blame each other for why their relationship didn't work, while children have the "broken home" syndrome as a model of struggle. In a Family DNA workshop we had a wonderful success story that demonstrated the power of clearing these imprints and the changes that come as a result.

I had asked the participants to become aware of a thought form of blame. One young man began the discussion about whom he blamed:

"You said that no one teaches us to not blame. I was taught the opposite. It was drilled into me that I should blame my dad.

"The thought form was that since he was never around, since he was always gone, that he made the

rest of our lives miserable. After my parents had been divorced for 5 or 6 years, when I was 15 or 16, I started becoming exactly like him and playing his role—the only role I ever saw him play, which was being aggressive and violent when he was home. I did it through drinking, like he did. I felt really guilty about being like him when I was in my younger, teenage life, but I couldn't help from blasting holes in the wall, raging, freaking out all the time and being out of control.

"Then, I started blaming my mom for not taking a stand and leaving my dad when things got really bad. It's funny because a month ago she said that she never left [him] because of us kids.

"Then I felt, 'No, I wanted you to leave, I wanted you to have a good life.'

"A thought form I had was in the image of a knot, just like two ends and this big thing in the middle of the struggle between them and back and forth sort of stuff. They both wanted to be free and they couldn't do it. I guess I was the knot."

Again and again I have heard children say, "I wish my parents had not put us all through their wars." They would rather their parents had separated and they hadn't had the fighting, hostility and the negative environment all the time.

"Though she said she stayed in the relationship for you, that may have been an inherited judgment that made her feel guilty, as if she would have been a poor mother by not staying. Especially since she came from a Catholic background, she might have lost God as well as her husband."

A more insidious reason people stay in poisonous relationships is the fear of financial ruin. We have become so dependent on dual family incomes that we have either overspent or are heavily invested in a lifestyle that feigns a higher status and we are loath to give it up. A mother may also feel emotionally as well as financially unable to deal with her children alone. Heart energy is the greatest wealth. Without it, no amount of finances will make life rich.

"Underneath all of your cultural and family inheritance, deep inside, there is another reason your mother stayed. The other reason is that she was working something out with your father. Sometimes, we want to keep up the fight because as long as we're fighting, we are still related. People stay together because of karma, because there are spiritual lessons they are learning from each other."

"I can feel that there is still something very deep between them."

The beauty of this story is that within a month of clearing his family DNA, the young man's father decided he needed to change his life and stop running away from his problems and his children. The parents actually spoke to each other for the first time in twenty years without fighting. The father came to visit and all parties have let go of the need to blame and are working for new solutions. The young man thought it a miracle. I think it an example of the power of dissolving negative inheritance!

I am sure you have had your moments of blaming, too. What were your dramatic thought forms?

Try this simple question: "Who do I blame in my

family?"

Close your eyes and see who comes up.

Ask, "Why do I blame them?"

You may see a picture or feel a memory flash into your consciousness.

Ask your body where you are holding the blame. Wherever you perceive the blame in your body, take your consciousness there and feel the energy.

Ask your body what color it needs to completely wash away the blame.

Draw that color into that place and let it dissolve the roots of the blame. When you can feel that all the residue is gone, take a deep breath.

Ask the person you blamed what color they need from you to come into harmony in your life and be released.

Imagine that you draw the color into the top of your head from the cosmos, bring it down into your stomach and radiate it out to them. See where they take it in. When they are full, imagine that they dissolve. Open your eyes.

This simple "exercise in consciousness" will begin a process of dissolving old emotional residues from your body so that both you and those you have blamed can discover what it is to be *free* from the burden of negative bondage.

Sending color to someone is a powerful way of becoming the "giver." Old patterns of the giver are laden with thought forms that connote being the giver is to be the martyr or forced to give. Most people feel depleted by giving because they do not understand the energetic principles of giving. For them, giving is merely a way of creating a debt, which someone will then owe to them. Actually, the giver is the one who has the power in a relationship because they are sourcing the energy flowing out to others. They choose to give. True freedom and the gateway to bliss is the ability to give freely—without counting how much has been given.

FAMILY CONSTELLATIONS

Our family, including distant relatives and ancestors, is a collection of beings who are genetically grouped together to form a constellation. We are intertwined by some mysterious cosmic force that we could call heredity, karma, divine will and even LOVE.

Intensives on Family DNA at The Light Institute include making a map of our family constellation. (*See illustrations, pp.64-65.*) The drawing helps us to clearly see how we psychically and emotionally relate to them and how we energetically group them around us. As we compose the map, we can describe our family constellation through a spatial relationship. We may feel closer to a grandparent or cousin than we do to our parents or siblings and so we mark them next to us. We might place someone out to the side of the page because we are disconnected or estranged from them.

People are often surprised to see the results of

their drawing. When the various family members are placed in front of them, they are able to see correlations of feeling, as well as the physical relationship. They become aware of something inexplicable and deep that causes them to be especially attracted to one or another of their relatives, without knowing why. These energetics come from the domain of the spiritual DNA.

There might be clusters of relatives corresponding to one side or the other of the family, or simply grouped by individual families within the constellation. When we see the nuclear family, the extended family and the relatives by marriage, we begin to realize how many people are connected to our lives—not by coincidence, but by some profound relativity that embraces us all.

It is important to include relatives who married into your family and even stepchildren or stepparents. You may find that you feel very connected to someone who is not actually related to you by blood. Sometimes that person even looks similar to you or has characteristics that you identify as your own. An aunt by marriage, for example, may share some qualities that are like your grandmother's, which you have inherited. As men often marry women who either look or act like their mothers, there is a bond of familiarity between you and your aunt. This is a circumstance that demonstrates how form expresses subtle essence.

I have come to feel that often stepparents are brought into a family because of some deep relationship to one or more of the children, as much as the one they have to the parent. The intensity may revolve around themes

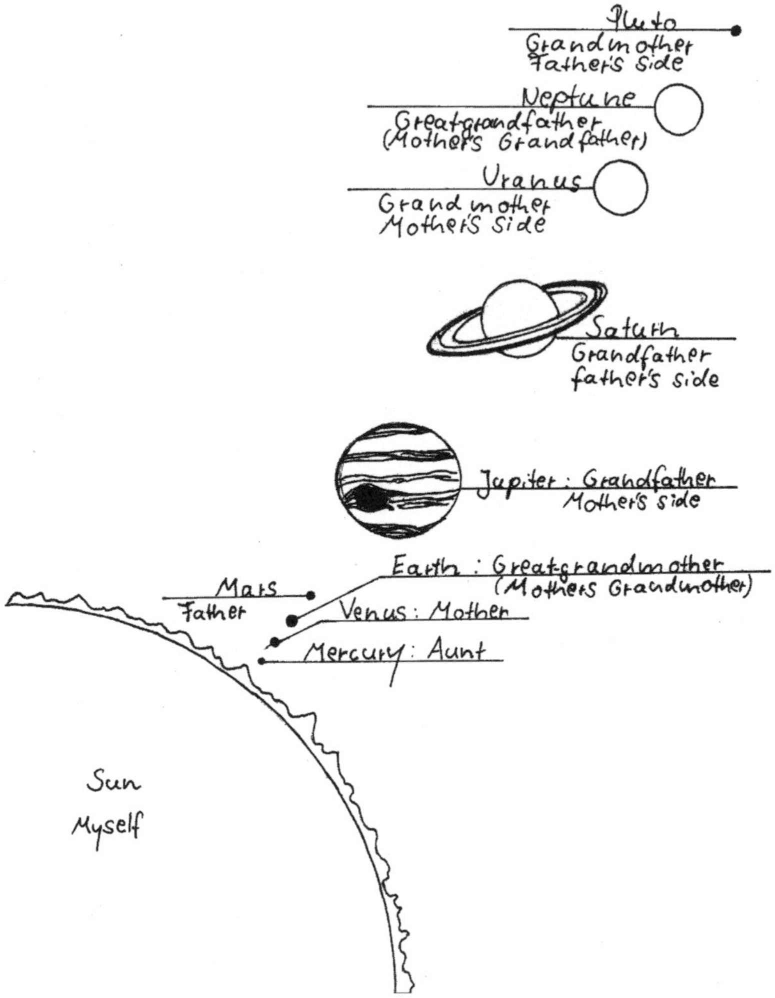

Family Constellation
Illustration 1

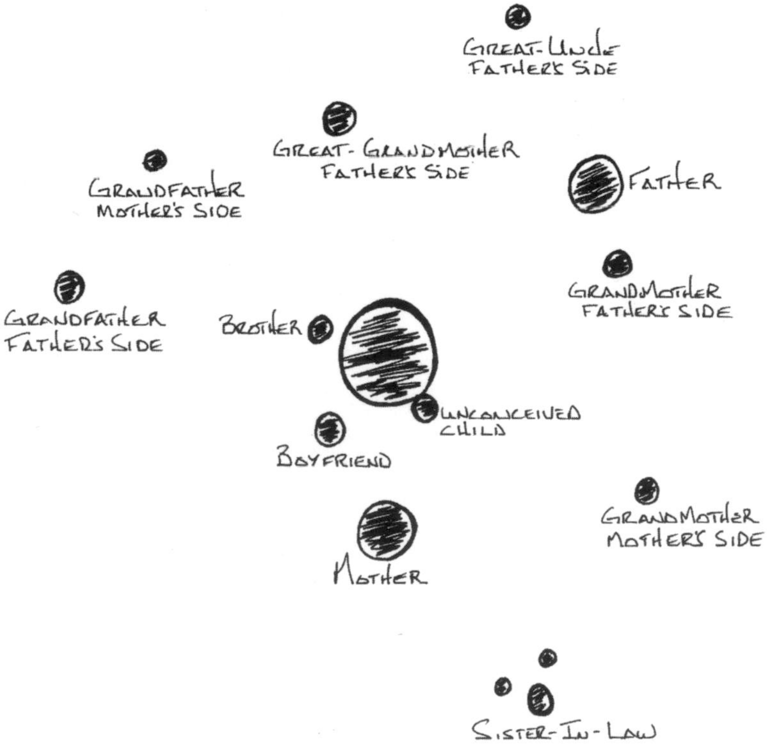

Family Constellation
Illustration 2

of jealousy or projection and may run the gamut from hate through friendship to great love. Whatever the arena, they are never accidental to one's lessons.

It would be very illuminating for you to make your own family constellation map so that as we move into the discussion of inheritance, you will know exactly what is true for you and who is most influential in your life.

To make your map, you can start by putting a dot that represents you in the middle of a blank page. Then begin filling in the family members that come to mind first. Though they are usually the immediate family, it may not be that way for you. Simply put the ones who are the most important to you closest to your dot.

You may have to concentrate in order to remember some of your extended family. Even if you never have met them, you should place them on your map because they represent your Ancestral Chi.

Another way to express your family constellation is through astrology. You could use the solar system or the galaxy as a structural format. Think of your family as a celestial configuration of beings who are grouped together and orbit around each other in designated patterns of relationship, like the sun and its planets. You could even assign family members to certain planets according to the energies of each planet as it relates to the predominant characteristics of that person. Thus, one person with a strong personality might be Mars, and another who has an outgoing, happy presence might represent Jupiter.

It is not unusual for there to be pairings or triad groupings between members in a family who feel much closer and understand one another more than they do others. This is true in the general public as well. Though we are all aware of our preferences, it may not have occurred to us to contemplate the source of our affinity to others. I am positively certain there is a shared inheritance that makes us feel so familiar to each other, whether is comes from our small, earthly gene pool, or our absolute Soul.

DIRECTIONAL INHERITANCE

There are several main pathways of inheritance within family constellations that facilitate awareness of our relationships to the various members of the family. By viewing relatives along their corresponding pathways, we can more clearly understand what we share with them in terms of heredity and karma. The pathways are directional and form vertical, horizontal and diagonal angles. (*See illustration, p. 75.*)

Parent-to-child and grandparents-to-grandchild lineages characterize the vertical pathway. Siblings and first cousins share a horizontal relationship. The diagonal pathway is the lineage from aunt and uncle to niece or nephew, as well as second cousins.

These directional paths of inheritance are especially illuminating in terms of the predominant exchanges, beyond the basic ones, that they offer each of us. You can imagine them as if you were looking at someone from certain angles. They might be taller or smaller than you. You might be seeing them face to face or from a side angle. Each perspective allows you to relate to them in a different way.

THE VERTICAL PATHWAY

The vertical pathway of inheritance is the direct pulse of life passing down from generation to generation. As it skips from grandparent to grandchild, a different dynamic emerges that reaches into the etheric realms and links us to deep ancestral and spiritual channels.

The vertical, blood relationship between parent and child sets up a huge mirror in which to mimic the self. The parents see themselves in their children and project their emotional repertoire of associations and conclusions upon the child. This is especially true between parents and their children of the same sex—between mother and daughter, father and son. In the same way that the daughter learns how to be a woman from her mother, the mother re-enacts her own childhood inheritance upon her daughter. Whatever thought forms she inherited about womanhood, relationship and sexuality are passed on at the moment of conception through her ovum.

During Light Institute sessions on the theme of clearing the parents, we bring the person's awareness to the moment of conception and view the thought forms of each of the parents. Your mother might have been thinking about how the dinner went or what she would do tomorrow. Beneath her lack of presence might have been a thought form about sex as a duty or emotional estrangement from her husband.

The mind of the cell is a perfect computer that imprints virtually every particle of experiential existence and sends it into the future through each new generation of cells. It seems unfathomable that what was going on in our parent's minds or bodies could be re-playing itself in ours, but it most probably is.

From the moment the fertilized egg attaches to the uterine wall, it begins to draw in nourishment from its mother. This directional pattern of pulling energy towards the self is the cause of much later suffering because the child pulls in the feelings and attitudes from his external surroundings and experiences them as his own, just as he did in the womb. The physiological body reacts to the external energies with physical symptoms such as headaches and stomachaches that come from the pollution picked up from others.

These emotional energies become a kind of negative environmental inheritance that causes us to feel the fear and anger of others without realizing that the feelings are not actually ours. We act out or identify with a myriad of precepts that we have taken in since before birth. This is why we find it is so very difficult to create distance between ourselves and our parents. Even when we promise ourselves that we won't become like them, we often do.

For their part, your mother or father see you as an echo of themselves and cannot understand your seeming rebellion to becoming who and what they are. The echo effect can be particularly difficult if your parents have their own issues about self-worth or feel that they must protect you from anything that seems close to their own experiences.

THE HORIZONTAL PATHWAY

The horizontal pathway of inheritance is filled with powerful projections onto our siblings and cousins that influence how we see ourselves. The qualities we perceive in others can become activated in us by associated inheritance. We learn from what is modeled to us and imprint it, regardless of who teaches it to us.

If they are older than we are, we often see them as more powerful or knowledgeable and so we try to emulate them. If they are younger, we may feel more threatened by them, as if they might take away what we have. These thought forms are usually instilled by others who are passing on cultural imprints. Behind the veil of competition for parental love and other familial recognition lies a profound connection through which siblings recognize each other as Soul friends. Even though they may fight, siblings are often very protective of one another to outsiders. It is not uncommon for a child to be aware that the mother is expecting another child before the mother knows of it. Two of my children announced the arrival of their younger sibling before I knew of it. When the spirit of the next baby contacts the parents, the child feels its presence directly.

In one of my sessions to uncover emotional DNA, I saw my sister staring up at the moon. I could feel her looking out into space and I knew that she was calling to me, waiting for me to be born. It was like witnessing the deepest of love. Something within the core of my being stirred as I recognized the magnificent bond between us.

Cousins often hold a special place in our hearts.

They resemble us in mysterious and fascinating ways and mirror family traits so that we can observe our inheritance, one step removed. As children, our cousins provide a safe place to practice relationship. We pretend to be "partners" and re-enact how we see our parents relate to each other. Sometimes these games include sexual overtures that are easier to explore with cousins than siblings. Countless people have laughingly confessed that they were in love with their cousins until they grew old enough to know that it was "verboten!"

When I asked my Higher Self to show me someone in my family from whom I inherited a particular trait, my cousin Dickie came into mind. I saw him holding a Zen stick and looking very determined.

I perceived the Zen energy in the way he furrowed his brow. It seemed to flare out energetically from between his eyebrows. I also saw that our eyes have the same shape and there is a similarity of countenance that radiates from our eyes.

In a flash, I realized that I was very much like him and that we both used our will to uphold what we knew was right. I remembered how as children I often felt that he hit me, psychically, to make me do things in the disciplined way he saw as correct. Many of our games held these components of him pretending to be a strict teacher. It is interesting to note that he discovered Zen Buddhism in his early twenties and has been a strict follower of that path for more than thirty years.

I can feel that same Zen energy in me and I can see that I have passed it on to one of my sons. Deep inside me, I am sure that it was something we did together in another lifetime and that is why we re-enacted it in our childhood in this lifetime. Now that I am aware of it, I feel that I can temper it with more love. It isn't enough to have the "knowing." It must be put into practice and expressed with gentleness.

Notice how a shared physical attribute (such as the shape of our eyes) is interwoven with emotional and spiritual aspects. In actuality, they are never separated: In the past, we simply have not investigated their relationship.

THE DIAGONAL PATH

The diagonal path from aunts and uncles is usually a supportive relationship. One step horizontally from the parents and then diagonally to you brings with it a capacity to instill more subtle genetic qualities than the direct blood connection. They may be character tendencies or talents, emotional perspectives or physical predilections of a gentler nature. While mothers and fathers are sometimes so close that they cannot distinguish us from themselves, diagonal relatives can more easily appreciate who we are and encourage and instruct us in ways our parents may not. Because they are not so invested in seeing themselves *through* us, they can actually *see* us. They may see something in a niece or nephew and help to bring it out through their loving support. Diagonals transmit inheritance through their words or touch and because of the blood connection, there is a vibration that occurs between the two that whispers the deepest of connections.

Think of your aunts and uncles and you will probably feel the warm sensation of the love they are willing to give you. They may be as critical of their own children as your parents are of you, but they heap praise and affection on you. In general, diagonal relations hold a sense of loving concern and are interested in their siblings' children. Contemplate whether you have witnessed your parents doing the same thing with your cousins.

It is also true that uncles are among the majority of molesters in the familial setting. It is very common for gay men to have experienced sexual fondling from their uncles, often at an early age. Girls, too, have often been at the mercy of lecherous uncles. Sexual hunger for nieces and nephews may be related to covert sexual feelings for siblings and even parents. Whatever the hidden provocation, there is a powerful attraction that stems from the blood level, which leads people to commit these acts that they might not perpetrate on strangers.

In many single parent families, aunts and uncles play a vital role in filling the space of the missing parent. This can be a very symbiotic arrangement for people who have no children of their own. They are wonderful at being the role model for a father or mother who is not there. Without the daily pressures of household survival, they add a comforting, loving element that both the parent and the children appreciate.

It is most mystifying to discover that you can pick up character traits and predispositions from aunts and uncles who are not related to you physically, but rather through marriage! Initially, you might think it has to do with learned inheritance or what has been ingrained in you through familial or cultural patterns. In fact, you have inherited them because of the psychogenetic pathways that link Soul friends together through shared incarnations and Soul connections that go far beyond this one life experience.

Though we consciously choose our mates from very specific matching traits, we are psychogenetically selecting their entire family constellation to include in our own, in order to set up and amplify the pathways of inheritance available to us. Although we are often surprised to discover how connected we are to certain members of the family who are not our direct bloodline, we experience that our feelings are true.

Whether hereditary links are vertical, horizontal or diagonal, awareness of the importance of our familial bonds is fascinating and an excellent way of reviewing aspects of self that we want to amplify or discard.

Directional Inheritance

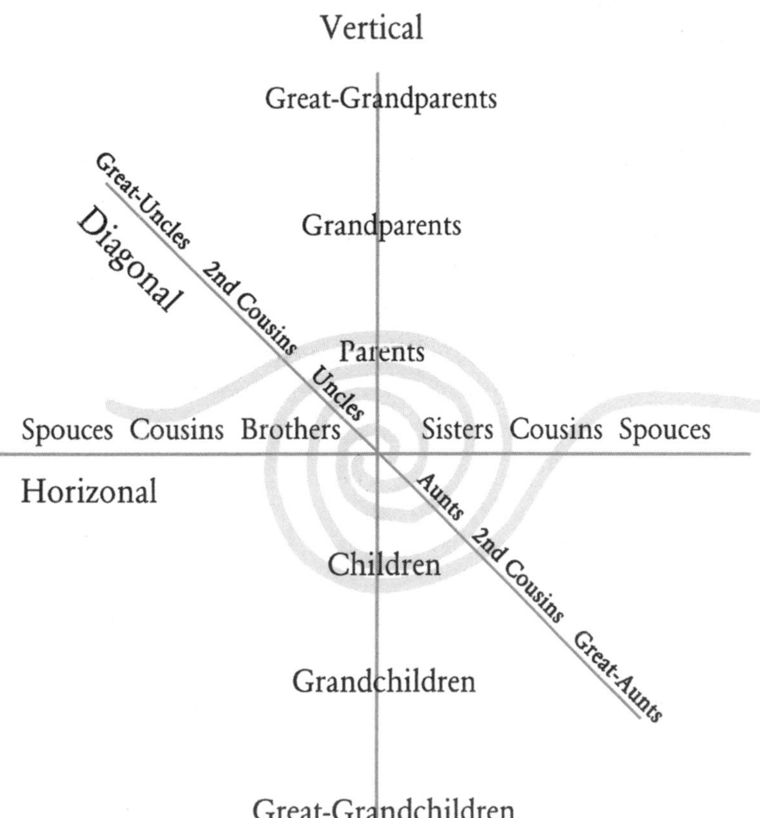

Illustration 3

FAMILY DNA

Our family inheritance includes emotional DNA and spiritual DNA as well as physical DNA. We can actually discover what we have inherited and from whom. When we ask our bodies to show us these traits, we can witness them in action as they have influenced our lives. The playback experience is brilliant, astounding and utterly fascinating. People we hardly ever think of or whom we've never met turn out to be the molders of our form and the instigators of our psyches. The entire panorama of family parades itself before our unsuspecting minds and rearranges all our fantasies and illusions, demanding that we re-stack the building blocks of our former familial constructs.

Let's look at some examples of inheritance that have been uncovered in Light Institute sessions on "Family DNA." These vignettes demonstrate the profound clarity that can be revealed through exploration of our family ties.

We begin with "progressed emotional explorations" that allow us to contact memories and experiences, pinpointing the hereditary qualities. In the first part of the exercise, we simply ask the Higher Self to show us a physical, emotional or spiritual trait we have inherited. In the second part, we look for a memory or experience in this lifetime that exemplifies that heredity. In later chapters, we will apply a similar format to discover how multi-incarnational exploration can illuminate the deepest hereditary patterns and how we can actually clear them from the physical, emotional and spiritual DNA.

PHYSICAL PATTERNING

"Ask your Higher Self to show you a person in your family constellation from whom you have inherited a physical pattern that you need to clear now."

From a young man:

"I see my dad. He has a big belly and I started getting mine when I was a kid. It's my stomach. I've always had trouble digesting—just like my dad."

"Ask your body to take you back to when you first began having trouble digesting."
"I'm a baby. My mother is giving me milk in a bottle. It doesn't feel good. It's too heavy. It hurts my stomach. I cry and cry. The more I cry, the more she gives me."

"See what else is there. What about your dad?"
"I can feel my dad. His stomach is very tight. I'm three years old. When he talks to my mother, I can feel how he has knots in his stomach. He has a lot of padding over his stomach to protect him. I see that I hold my stomach the same way. Food sits in my stomach, just like in his."

"Ask your stomach what color it needs to release the tightness so that it can relax and digest."
"It needs green."

"Draw the green, like liquid light, into your stomach and see what happens."
"My stomach is starting to get soft. I feel the green coming in through the umbilical cord."

"Ask your body where it is holding the experience of

crying when your mother is feeding you the heavy milk."
"It's in my chest."

"Ask your chest what color it needs to erase this memory, now."
"It needs pink."

"Draw the pink into your chest and let it wash away that experience. Tell me when it is gone."
"It's gone. I feel very open."

"You can actually send healing energy to your father to help him, too."
"I'd like that."

"Ask your father what color he needs from you to heal his indigestion."
"Blue."

"Imagine that you are bringing in a blue light from the cosmos, into the top of your head and down into your stomach. Laser the blue light towards your father and see where he takes it in."
"It comes in through his stomach."

"Good. Keep sending it to him until he disappears."
"He's gone. This is amazing. I feel great!"

EMOTIONAL PATTERNING

From a woman in her early forties:

"Ask your Inner Child to show you a person in your

family constellation from whom you have inherited an emotional pattern that you need to clear now."
"It's my Aunt Betsy. She always pretends that everyone is yelling at her and that she is just mirroring their anger when she yells back in self-defense. It's self-righteousness. She thinks she is the good guy and everybody else is trying to control her."

"Ask your Inner Child to take you back to a time in this lifetime when you experienced this same kind of self-righteousness."
"Just last week."

"What happened?"
"My mother was telling me that I should have done something for my grandmother. I saw her as manipulating everybody around her, so I told her off. I felt sick of her always telling me how I should behave when it was she who was being such an angry jerk. Maybe I was being self-righteous at seeing her as the one who was yelling. I guess I was yelling too."

"Ask your body where it is holding this memory from last week."
"It's in my throat. It feels very tight and dry."

"Ask your throat what color it needs to dissolve the memory with all its feelings."
"Yellow."

"Draw the yellow into your throat and wash away that whole scene. Keep doing it until your throat feels changed."
"Wow! It feels really open. I feel like I could sing."

SPIRITUAL PATTERNING

"Ask your Higher Self to show you a person in your family constellation from whom you have inherited a spiritual pattern that you need to clear now."
"I don't know this person. It's way back on the paternal side. I think it is a great, great uncle. It feels like he's in Ireland. Red hair. Now I can see him very clearly.

"He's extremely spiritual, but terrified to show it or any kind of power he has. It scares him. He's good-natured but he has this deep, underlying spirituality that he can't share with anybody. He's very afraid to let people see that."

"What is he afraid will happen if people see it?"
"They would see him as weak, or delusional to think that the power of God is inside him. He feels like it is something that would make people think he is feminine or something. It's somehow against his normal religion. It is like a connection to the cosmos. It's something deep inside him that he is hiding. It is so frightening because he doesn't really know what it is. It's a beautiful, spiritual knowing."

"I inherited this fear of the beauty of spirituality."

"Ask your Inner Child to take you back to a memory in this lifetime when this inherited fear came up."
"I am very young, about four years old, and I hear people talking about religion and church. I know that what they are saying is not right: this angry God business—how God's gonna punish you and you have to live by this...always being afraid of God. It feels like such brightness inside of me to be able to know that

God doesn't punish, but I think there's something wrong with me for feeling this way. I can't tell them. I don't know how."

"Where are you holding the residue of this fear of showing your spiritual knowing?
"In my genitals."

"What color does it need to be washed away?"
"Cobalt blue."

"Draw that color directly into your genitals and let it absorb all the fear until it is gone."
"It's gone."

"Where are you holding the memory of having spiritual knowing inside you that you need to hide?"
"Both of my knees. They need red light."

"Allow that red to free your knees and give you the strength to not hide anything you know."
"They feel light. This could change my whole life!"

It is most illuminating to peruse your family constellation to see what you have inherited or what you share with the various members. Once you become conscious of these traits, you will be able to use them or discard them in your life.

Here is an exercise in consciousness to help you pick out the karmic links with your relatives, whether they are physical, emotional or spiritual in nature.

Read through the exercise several times and then begin.

(Your *Higher Self* represents your own inner voice or intuition. Just ask the questions and accept what comes into your consciousness.)

Close your eyes and breathe deeply into your body three times.

Ask your Higher Self to select someone within your family constellation with whom you have a karmic bond. You may get a picture of the person or hear their name.

Ask to be shown something you have inherited from or with this person.

> It might be a physical trait such as the shape of your mouth, the form of your body, or a tendency to nervousness or colds.
>
> It might be an emotional characteristic such as temper, the need to control, or a friendly disposition.
>
> It might be a spiritual quality such as a creative gift, a love of nature, or a contemplative lifestyle.

Once you are aware of the inheritance, ask your Higher Self to show you an experience or memory in which you used that inheritance in this lifetime.

Ask your body where you are holding the memory.

Ask what color that part of your body needs to release the memory.

Draw the color into that part of your body and dissolve the memory.

When you feel that it has been freed, ask that person what color they need from you to come into balance in your life now.

Imagine that you draw the color from the cosmos, into the top of your head, down into your solar plexus, and radiate it out to them. Imagine that you perceive where they take it in. When you feel they are full, give them a little more color until they disappear.

Take a deep breath and open your eyes.

The results of this exercise are often astounding in terms of the information they offer. It is not uncommon for your Higher Self to show you something you inherited from a relative you never think of. You could actually go through your entire family constellation to find out what you share genetically with each person. In so doing, you are loosening the fabric of your genetic blueprint so that you can re-weave it in a new way.

It might be especially illuminating to check the links with anyone you do not like in your family constellation. The people you least like are often the ones with whom you most closely share specific traits. Some of the psychogenetic residues or reference points are actually sourced in other incarnations which your family members have shared with you and therefore carry forward to the beginning of your relationship in this lifetime.

The emotional qualities of projection play a role in our feelings about others: We perceive negative attributes in them that are actually inside us. Although we cannot fathom why they are so irritating to us, we are vulnerable to like frequencies that reside in our own psyche.

Chapter 5

PARENTS:
THE BLOOD CHI

All families have a sense of communion that is transmitted through the blood. "Blood of my blood" is a bond so strong that almost nothing can release it. Whether we are angry with our parents or are very attached to them, there is something that holds us together, even if we do not see them. As time goes on, we become more like our parents, both in features and in life perspective, because the seeds of childhood teaching begin to sprout. Even after they have passed on, we hold them. Carrying their blood in us creates an indelible mark in our being. The blood Chi is an irrevocable signature of relationship that cannot be erased; it can only be mastered through unconditional love.

It may seem a long journey from childhood to adulthood in the fenced environment of our parental enclosures, and it is an adventure laden with dead-ends and infinite skies. The liquid path of our blood takes us through each defining moment experienced by our parents and the corresponding result as it fixes itself to the DNA that whispers sameness to our own cells. We re-enact it all and then wonder who we are!

Without the profound benefit of consciousness, we end up closely replicating our parents. When we are young, the strong force of rebellion and the need to find ourselves, coupled with our physical energies, protect us and guide us on our own path. Later on, as we settle into our lives, the more subtle imprints surface and long-standing parental reflections begin to emerge. The Soul agreements that brought us together can be seen through their infiltration into the body. As we grow older, we take on the facial and bodily features of our parents more strongly, as well as their philosophical or personality quirks.

It is essential that our old karmic patterns be resolved, rather than continued. To release the genetic bonds, we must come to recognize our parents as our closest Soul friends. We know before birth if they will be present in our lives and whether it is going to be a difficult lesson. Our parents are here to help us free ourselves from psychogenetic residues that we are ready to transmute and release.

Parents are our greatest gift because they give us life. They are the vehicles that lend passage to the Soul's expression. Sometimes they do it "accidentally," without consciously wanting a child. Sometimes it is their deepest longing. Whatever their intentions, conception is a result of the unfathomable Soul, clamoring to take form. Life wills itself upon the primordial body, which gladly complies, regardless of the consequences.

These divine accidents never occur between strangers; they are always agreed upon by Soul friends who are willing to be the teachers, enemies or accomplices of each other in an endless round of give and take. Here in the constriction of the third dimension, it is sometimes difficult for us to comprehend how we could have chosen each other, but we have. The choice has always been from the level of the Soul and always for our highest good, though sometimes it feels more like punishment or revenge.

It is not only that we have unfinished business, but that we do each other favors by playing parts, albeit unsavory parts, that no others are willing to play. Especially today, we are surrounded by psychological banter meant to support our justification that we are less than perfect because of the terrible tarnishment of our parents; who, of course, are equally adamant that it is all caused by their parents.

Our experiences in childhood at the hands of parents and others do not give us cause to be less than our true selves; rather, they are the spiritual lessons we have come to grow through. They challenge us to find new answers and solutions to our deepest dilemmas. Blame is a great impediment to growth. It is much better to contemplate the nature of the gift provided us by all experiences, especially since our attachments become our DNA and therefore, our future!

It is not only in this lifetime that we latch onto those we call parents; they have been our lovers and friends in many sojourns into body. Although the triad of mother, father, and child is a perfectly balanced composition, it is often two of the three who are psychogenetically-linked and the third might be a more distant participant. One of the parents may have called in the spirit of the child to work together on a theme that corresponds specifically to them.

Soul beings will often construct a kind of etheric web or pre-substance to link them to a possible parent with whom they have some karmic connection and who is perfect for their life's lessons. I have witnessed this energy lingering around the shoulders of only one of the perspective parents and noticed that the kindred spirit may accompany its chosen parent for some time before impregnation actually occurs. This is true for men as well as women. There are always strong vows or contracts from other incarnations that predispose this arrangement. Whether the chosen parent changes mates does not seem to matter. Even if an abortion occurs, the spirit will wait in the astral plane for another opportunity to incarnate with that Soul partner. This may be because the parent is sending mixed messages such as, "Not now, maybe later."

My friend Lea Sanders, who was one of the world's auric masters, used to talk about children coming in on the "ray " of one or the other parent. By this she meant that there was a certain resonance between the baby and the parent that came from shared emotional or spiritual DNA. They held specific frequencies in common that would allow them to understand each other and shared qualities that could make them appear very much alike.

A child who has chosen a parent for certain karmic reasons might look or act more like that parent than the other. It is as if the entire physical, emotional and spiritual DNA were more intertwined and homogeneous between them. The depth of their relationship reflects itself in their oneness on all levels. They may behave more like a couple who has known each other forever than a parent and child. This usually occurs between parents and children of the opposite sex, though it could happen between mothers and daughters or fathers and sons. In spiritual terms, the harmonic energies began long before this incarnation and the bond between them is evident to all onlookers, although they could not envision the cause without including awareness of spiritual relationship.

The heat of the blood is a special bond that causes profound confusion in relationships that should follow a prescribed format, such as parent and child, but which actually play out entirely differently. For example, a child may take the role of parent to the parent or even carry on like a lover. Without the background awareness of their relationship, parents and children cannot possibly conceive of why they feel the way they do about each other or even allow themselves to know their feelings.

So much sexual and emotional guilt could be released if the world understood that parents and their children have known each other in many intimate arenas. This spiritual knowledge could possibly eliminate a great deal of "child abuse," freeing these powerful unresolved physical and emotional seeds to find other channels of growth.

The intimacy between one parent and a given child may seem exclusive of the other members of the family. Often, this special relationship triggers jealousy and hurt feelings in the parent and siblings who are "left out." All too often, a mother and daughter engage in a devastating rivalry for the affections of the father without knowing what is really happening. Father and son may be equally as competitive for the attention of the mother.

How do you feel about each of your parents? The contracts and vows you hold with each of them individually cast the mold of your relationships, which cannot possibly be the same for both of them.

Ask your Higher Self to show you a contract or vow you have made with a parent. You will be amazed at the answer. When you do sessions, you can see exactly why and how you made those contracts. They may come from another lifetime in which you vowed to never let go when a love was taken from you, or a vow of vengeance to get back at someone who overpowered you. You will carry over into this lifetime those psychogenetic conversations and rebalance them in accordance with your Soul's mandate.

All of us as children intuitively know what is happening to our parents and want to help them. We say, "Let me do it. Let me make it OK for you." The way we do it is usually through our bodies. We suck negative energies in through our solar plexus, umbilical area. You may have promised to take care of one of your parents and find yourself doing so from early childhood.

Parents also experience a sense of vague contracts or agreements with their children, even before they are born. Mothers tell us that their children's personalities are felt in the womb. They often have a sense of defined relationship from the beginning that is different from other pregnancies. Some little kickers are rambunctious right from the start, while others *in utero* are very subdued and even experienced as tender by their mothers. If we could recognize that they have a pre-established relationship, we could see the subjective responses already in play between them.

Sometimes they have a profound rapport and the mother feels she can talk to the fetus as if it were someone she already knew. Other times there is a strange feeling of some foreign energy in the body. All these feelings are part of the channels of relation that draw us together, as well as the specific karma between the mother and each child.

Fathers also have special connections with each child. The energies that pass between them while the child is in the womb are truly psychogenetic. From the reaches of their blood, they communicate psychically through genetic pathways that link them together, even though they may not be consciously aware of it.

The mystery of inheritance is fascinatingly played out in the situation in which parents and children have not seen each other physically. When we've never seen a parent, it seems as if we reach deeper into our psychic matrix to connect with them and somehow are able to enact their attitudes and characteristics with even more fidelity.

Many mothers have confided in me how maddening it is to watch a child who has not lived with its father take on attributes that describe him perfectly. This mirroring effect is so disturbing to the mother because those qualities may be the epitome of what went wrong in her relationship with the father in the first place.

As the child unwittingly plays out past emotional

dramas of the mother, she fears that she will lose again. Her response is often to repeat the battle with her child, who cannot possibly understand what is upsetting its mother. Projecting her feelings about the father onto the child, the mother may engage in psychological warfare, misconstruing the child's actions and intentions as those of the father. That the child gestures or looks like the father is subconscious proof to her that the child is "on his side." The result is a child who grows up feeling that everything is his/her fault, without really knowing why.

Verbal attack is a temptation that must be avoided, as it is so destructive all the way around. Though a parent may feel justified, what we have created in our lives reflects our inner choices and cannot be rectified by attacking back. The more alienated one is on the outside, the more the inner psychogenetic bond will grow.

Because parents are the ultimate authority, whatever they say you *are*, you will somehow *become*. Their descriptions will, to some degree, echo down from the ones they heard as children from their parents and they will see you in the same light as they were seen themselves.

Stop for a moment and remember what your parents told you about *you*. Did they call you stupid, clumsy, brilliant, pretty? You probably took on those descriptions and made them fit. Think of the power of those labels; to spend a great deal of your life trying to match a word, not because it is your destiny, but because it makes you what your parents expect you to be.

When we are working on "Clearing the Parents" at

The Light Institute, we guide the person back to conscious awareness of the moment before conception. They look down upon the parents they have chosen to become cognizant of their innate qualities. The humor, the temper, physical constitution and mental capacities are all picked in pre-packaged arrangements to suit our needs. Why do we choose what we do? How do we do it? It is a legacy that spans lifetimes, our ancestors' and our own, weaving the strands of our DNA into the perfect expression of human purpose. What a twist of fate that we ourselves actually select our inheritance!

The thought forms and feelings of our parents during conception are very poignant for us. It is such a cosmic moment when the sperm pierces the egg. The egg extends a biochemical invitation to one particular sperm that carries exactly the genetic encoding desired by the egg for its perfect union. Their fusion anchors in a new life. How does the egg recognize the coordinates carried by the sperm? Since it calls to only one, it must already know its purpose. This is *conscious conception* and we cannot yet fathom what it is telling us about our holographic Souls. At that moment, we are choosing the perimeters of our future, the labyrinth of free choice.

Unaware of the ecstasy of the sperm and egg, our parents may be in another world of separation or despondency. Whatever their state, the internal dialogues are lifted verbatim and cemented into the psychogenetic material of the fetus. It is astounding to hear people describe the thought forms going on inside their parents while they engage in the act of conception.

Sexual fears and emotional tête-à-tête seed themselves into the new cells and await a not-too-distant future when they will be called into play by experiences that source back to them. Unwittingly, we replay those imprints in our own life experiences.

What we witness in viewing our parents allows an instantaneous shift in our emotional fields. By tuning into their inner feelings, we are able to perceive the deeper realities of our fathers and mothers in their own primordial soup. We can begin to see who they are and who they are not.

So often we felt unwanted or that we were the wrong sex and through this lens we discover that our conclusions were untrue. They may not have wanted a child, but that is very different than not wanting *us*! Great healings take place when we revisit these formative moments and see life from their perspective.

Our insatiable emotional bodies never want to give up the wanting. There is something almost delectable about the longing and the waiting for someone to give us our lives, which is exactly what our parents have done. Yet we focus on what we *didn't* get, rather than on how clever we were to have chosen these parents.

Birth itself is a formative moment in the psychogenetic process. We not only take on the emotional imprints our mothers are experiencing, but we inherit her conclusions.

At The Light Institute we help expectant parents to clear the birthing process. Imagine the imprints you have received from the lineage of women in your family and your husband's family who have had adverse birthing experiences, not because trouble is normal, but because their bodies have not been allowed to joyfully let nature take its course. Each difficulty is amplified as it is passed down through the psychogenetic pathways, while the odds of peaceful birthing diminish because of fear, which separates us from our body's wisdom.

It is crucial to release the inherited fear and negativity surrounding the miracle of birth so that the body can do what it knows how to do perfectly and reinstate birthing as a sacred initiation of the Soul.

During pregnancy and birth, our mothers formulated many impressions and responses, which we inherited. If she wanted her pregnancy to last forever, or was afraid to let go, we may very likely translate those energies into our philosophy of life. Many of us "hold on" for dear life without knowing why we are so tenacious.

All three members of the sacred triad may interpret birth as a point of separation. The mother separates from her child. I will never forget the indelible sensation of feeling as if I were lost in space when I myself cut the umbilical cord of my fifth child. I wanted to birth her and bring her into the world myself, as my own gesture of love, but I had not expected the sense of loss. It was as though I had cut off some part of me that could never be returned.

The child is separated from its wondrous womb and pushed out into the world. Fathers are often shut out from sharing the birthing experience, separating them from something they want so deeply to be a part of.

On another level, we enter the world encased in a body that soon separates us from conscious awareness of our divine selves. The amnesia is solely a transitioning mechanism into a focus of form, not meant to be a loss of source. Our parents are the vehicles of that transition which we ourselves have chosen.

Have you ever really thought about what it means to inherit your body from your parents? The "good" genes and the "bad" ones are only a tiny part of the conversation. How they feel about their bodies has been passed to you in the same way they have seeded their height and constitution. Their fears and dreams, as well as what they were taught about successful bodies, are inextricably folded into your psychogenetic blueprint. Through your consciousness, you can extend the most positive qualities to those who will take up your inheritance.

Perhaps life causes us to forget the perfection of selecting parents who would offer us exactly what we have determined to be the karmic lessons we need for the growth of our Soul. In Light Institute sessions we ask to look at the psychogenetic inheritance contributed by each parent. This brings us a powerful sense of intimate connection, which we may not have remembered under the bondage of emotional conflict.

Through the Family DNA sessions, we can decipher an ever-widening arc of psychogenetic repertoire that illuminates why we have chosen our parents and family. Whatever the contracts or vows, the DNA will reflect our choices and the subtle nuances of emotional and spiritual predilections will weave themselves into the genetic fabric.

Though what we see may be a likeness of hair texture or the shape of the mouth, these physical attributes house the emotional and spiritual energies that are expressed through the physical form. When bodily predilections arise, they are as likely to come from the karmic emotional and spiritual DNA residues we share, as from the physical constructs. The blood Chi is a river flowing home to the source of our oneness. Through our family, our culture, our species, the blood repeats the patterns of psychogenetic relationship.

Chapter 6

ADOPTIVE LINKS: THE ENERGIES BEHIND ADOPTION

There are special karmic principles at play between adopted children and their parents. The deep bond of love that links them is exceptionally strong, born from the circumstances of their union. Here is a couple who long so much to hold a child in their arms that they care nothing about the primordial blood to blood bondage. I have heard such powerful heart stories of parents who chose their child with a profound feeling of relationship without even knowing the child.

A child, given away and marked by the psychic stigma of adoption long before the mind can even comprehend the meaning of the words, knows that these are the right parents.

No illusions of accidental conception cloud the acknowledgement that the partners (parents and child) picked one another. The act of adoption itself is a careful process of cross checking to make sure there is a good match for the adoptive child and the new family. Though the child may not have the opportunity to say no on this level, the force of synchronicity orchestrates the circumstances so that the prospective parents are led to that child.

There is a sense of mutual commitment between them as if they knew from some far recess of their being that they had set up this agreement. On the spiritual level, there is a clear choice to come together. The psychogenetic magnetism draws the sacred triangle (mother, father, child) into the powerful configuration of interdependent support.

The connection is beyond physical happenstance. It is an agreement made by Soul friends who already know what gifts they will give each other. They take up their parent/child roles from a much loftier level, without the blood Chi to cloud true recognition of the gift of their relationship.

The Soul has a purpose in setting a being free from the web of family DNA. Even other siblings who are not blood relatives support a kind of freedom of relating beyond the suffocating "stickum" that is present in the contracts of blood families.

Adoptive links allow us to witness the subtle truth of psychogenetic inheritance because they begin where physical DNA leaves off. They provide a strong clue for the existence of non-physical heredity that shows itself to be environmental, emotional and spiritual in nature. This environmental inheritance, both spiritual and emotional, bleeds into the physical.

In a spectacular show of psychogenetic hereditary transference and intricacies, families with no blood Chi often look physically alike and take on each other's attributes by virtue of their shared psychogenetic environments. The emotional and spiritual DNA *bends* physical attributes into likeness by virtue of mimicking. Thus, the child develops a similar characteristic because of the uncanny ability to copy the gestures and poses of the surrounding family.

Since look-alikes confer a sense of belonging, children become brilliant at copying the bodies of their elders and siblings even without the blood patterns. Eventually, the mimicry will coalesce a biochemical signature that will redesign the DNA.

It may seem hard to believe that a child could influence body form by simply mimicking a parent or sibling. Yet, surely you have seen people around you, mothers and fathers, who begin to look alike after years of living together. It is the same principle. Perhaps it began with a physical signature; the way one raises an eyebrow, for example, and ultimately the shape of the eyes follows the gestures into a permanent likeness. The correlation here is easy to understand because the sphenoid bones that affect the size of the eyes are easily altered.

The same is actually true of the rest of the body as well. If you hold the body in a certain way, it will develop musculature that allows it to maintain that shape. Physicality is symbolic of emotional and psychic repertoire. We are all so much more alike than we can imagine.

I find that after a small amount of time, even animals begin to look like their owners, or vice versa. Not only do they resemble each other, but they adopt the personality traits of one another. This is perfectly plausible if we recognize the psychogenetic grid that links emotional frequencies to physical and spiritual ones, including the evolutionary chain of all species.

A small boy may admire the thick chest of his adopted father. He will imprint the vision within his inner reference points as to how he should look and be when he grows up. That stored repertoire will begin to influence his biochemistry until the genes and chromosomes have received the mandate and begin to organize themselves into DNA strands that produce exactly that effect.

Another small boy may look up at his blood father and feel that he will never be like him. Perhaps the father scolds him for not being more manly. His sinking feeling that he cannot meet his father's expectations will literally cause him to shrink his chest into a concavity that looks nothing like his biological template.

In Light Institute sessions on Family DNA, adopted people have had no trouble identifying physical traits that they inherited from their adopted relatives. They have shown me pictures that bear out the truth of those likenesses.

I have studied the facial similarities between adopted children who are even from a different race than their adoptive parents and yet look very much alike. One family I know has four adopted children from several different races and every one of them carries features that are almost exact copies of one or the other of their parents.

It may be that the parents were drawn to adopt a child from that race because of another lifetime of their own when they were of the same race or had a positive interaction with them. It is possible that they might carry over several isolated physical features characteristic of that group into this lifetime.

It is especially fascinating to me when I detect a characteristic of civilizations that left its mark on our global heritage and shows up within individual body structures. I, for example, carry the slanted forehead of the Atlantians that was passed to the Egyptians and then on to the Mayans. My short tibia lower legs and small shoulders are Asiatic and probably of Galactic origin. My high cheekbones are reflective of my Algonquian heritage, but the blue of my eyes is Nordic. As we learn more about human migrations, and the specific physical features of our races and origins, we will perceive clearly the history of our ancestors and the imprints of our many sojourns on planet Earth.

The more enlightened we become, the easier it will be for us to see the gift in the irreversible situations of our lives. When we can look down at conception and see that one set of Soul friends will provide the blood templates and another set will tender the love and environment for growth, it will seem perfectly natural and good that it be that way. Since all parties are related to each other through infinite incarnations of the Soul, there need not be confusion or abandonment. Indeed, much of the suffering about adoption occurs because we are caught in a shadow land from which we cannot see the Soul's purpose.

We have collectively interpreted embodiment as the punishment of separation. It is a universal theme of humanity in which we feel that the stigma of separation is stalking our every experience. At birth, we are taken away from the blood that nourishes us and the secure protection of the womb that holds our sense of inclusion. The truth is that separation is painfully experienced at birth, due to the amnesia of the Soul, regardless of adoptive or blood mothers. Almost all of us react from an abyss of confusion and uncertainty that clouds our first moments of life.

In the case of adoption, uncertainty inevitably becomes a deep-seated insecurity that battles the whispers of possible imperfection, guilt or unworthiness. No rational explanation could ever resolve the complexity of emotions that result from being given away. We can "understand" how it happened, but never why it happened—to us!

The answers to such profound questions are even more complex because they are sourced in the spiritual recesses of our Soul's choices. If we could move from the trauma of such an unanswerable act, to the acceptance that these surrogate parents are, in fact, the target of our programmed family matrix, adoption could be perceived as the exquisite love story that it actually is. There is such a gentle respect and appreciation between the Soul friends who have found each other in this way. Why is it that we are taught to focus on the drama of not being wanted instead of the gift of receiving energies from two sets of Soul friends who are giving us their best?

Society, as a whole, projects its collective separation trauma onto adoption with all the judgements concerning good mothering and acceptable behavior. If we choose to view adoption from the clarity of spiritual mandates, there would be no stigma and the negative effects would be minimal.

How many young girls have been counseled to

give up their babies because they were told that they surely could not mother them and that adoption would be best for the baby? In the end, the young woman must deny her own instinctual knowing and abdicate her potential to the authority who later holds her guilty for so doing. The result of not trusting ourselves is always an uneasy sense of insecurity, coupled with self-defacing feelings.

The insecurity of all concerned is profoundly increased as a result of the layers of secrecy that are placed upon the facts of adoption. Adoptive parents are sometimes afraid to tell their children the truth because they rationalize that it will be painful. Underneath all their reasons is an underlying fear that the child would reject them and try to return to the blood parents.

This scenario has little truth to it as children are deeply loyal to those who parent them—blood or no blood. If they understood the karmic perfection of their relationship to their adopted children from a spiritual perspective, they would realize that there is nothing to fear. The fear of rejection shared by both parties plays itself out again and again throughout life.

If the conversation of biological parents is never breached, adopted children may become obsessed with the fantasies of what their "true" parents are like. It is a primordial and appropriate quest to want to see what those who created our flesh, look like. It need never be seen as a lack of appreciation or love for the adopted parents. It is imperative they realize that it is not about them in any way.

There is much less karma intrinsic to adopted relationships than that which is spiraling around shared physical DNA. Yet the themes of insecurity that plague both parents and children are mirrored back and forth. The adoptive parents must wrestle with deep, primordial confusion about not being able to conceive, while the adopted child will feel the confusion of incoming messages bound in subtle psychic inflection that give them a vague sense of strangeness about themselves.

The psychogenetic pathways between us cannot be disrupted by absence alone. Thus, the thoughts or feelings of the blood parents are always accessible to the child whether or not s/he ever meets or reunites with them. The shadow of shame and guilt borne by the parent is felt by the child on the psychic level without the benefit of knowing that these negative energies belong to someone else. The adopted child often carries a sense of vague guilt as if somewhere in the unremembered past some unspeakable act had been committed. This is the result of the feelings and thought forms flowing in through genetic encoding, not any absolute truth. With psychogenetic clearing and emotional training, we can learn to transcend these misguided energies and come to know who we are.

As the invisible worlds unfold, all humans will begin to experience a sense of lineage that links us to yet a greater family of relations than the environmental or blood ones we now recognize.

Chapter 7

GRANDPARENTS: THE SPIRITUAL CHI

Most of us love our grandparents and feel loved by them. Whereas we perceive ourselves to be judged or punished or ignored by our parents, we almost never assign such negative energies to our grandparents. It is a basic societal concept that grandparents give unconditional love and can be trusted to adore us, as well as be a voice in our favor within the family. At least once in our lives, most of us have relished the situation in which our grandparents have defended or sided with us against our parents. Just the fact that someone has power over our parents is awe-inspiring, at least at some ages.

We admire our grandparents for their freedom and power over the constraints of others and their ability to disengage from the extraneous details that create struggle in life. We love them for listening to us and even joining us in our worlds from time to time.

The young and the old enter into "dream time" easily as they are more sensitized to the pull of invisible worlds. For the grandparent, it might be a memory triggered by some association that makes them revel in the re-experiencing of a past saga. It is common for an older person to remember in great detail an event that occurred forty years ago, but not be able to remember if they took their vitamins this morning. The remembrance is bound to the timeless astral dimension by the sounds, smells and images that flood their senses and replay life.

On the other hand, the grandparent may have begun the long passage of meltdown to the other side of the veil. As s/he disconnects from the world outside, it is easier to slip ahead into the energies that will carry them home to another kind of life, without a body. Neither a child nor an old one is very far away from these other worlds and so can recognize their energies that are not perceived consciously by the rest of the family. The grip of the cosmic currents whispers a pulse that brings them into worlds we have completely forgotten or have relegated to the domain of imagination.

Being in the early twilight of their lives while we were in the spring of our own, we and our grandparents have had many "inner world" experiences and secrets. Since we could not share them with our busy parents who were a source of validation, they have now become vague shadows of a childhood we cannot recall.

Children effortlessly float out of body or across the astral veil, through the fluidity of consciousness that excitedly follows any flicker by the beings that beckon from beyond. Grandparents and grandchildren are transfixed by the beauty of their dream worlds that take up the space and buffer them from the fast-moving outer world. They are best attuned to the slower alpha brain frequencies that allow more creative yet deeper awareness.

Because the blood Chi is one step removed by the intermediate generation of the parents, the emotional interference is likewise greatly lessened. The grandparent is more willing to enjoy the child just as s/he is and is able to pass down the wondrous energy of the *spiritual Chi*.

Perhaps it is that grandparents have tired of nudging their own children to do what they deem proper and right. They are less interested in the rules of society, having learned that all the fuss is not so important in the end. They are infinitely more patient and accepting and can now freely bestow on their grandchildren the love that they could not give their own children.

Sometimes the grandparent actually learns how to love from the grandchild, who will help thread the love back to the parent caught in the middle of the family relationship theme—child of the grandparent, parent of the grandchild. It is an intense point of karma as the two poles both use the middle to work off their own sense of self.

The spiritual DNA encompasses the interlocked physical and emotional DNA and becomes the point of closest proximity to the inheritance of the grandchild. By its very nature, it allows contact with cosmic energies as well as the almost infinite flow of spiritual experiences passed down through the ancestral Chi.

Whether it is ever discussed or alluded to at all,

the grandparent will hold the spin point of the spiritual dimensions, allowing the grandchild access through the spiritual DNA received from the "elder." It is not what the grandparent does on a conscious level in the daily process of life; it is the dreaming and sleeping wherein they travel out and back across the veil that teaches the child to cross over too, and anchor in the experiences.

You may be thinking about your grandparents, who perhaps seem non-spiritual in the traditional sense. It is a bit of cosmic humor that they may cuss and swear, defy "God," or live anything but a spiritual life and that it matters not! A grumpy and inconsiderate grandfather could possibly pass down a powerful spiritual inheritance to you by virtue of his spiritual Chi. That has nothing to do with his choice to act on this energy in his own life experience. He may have no awareness of it himself, nor have any interest in spiritual conversations and yet become the source of divine flow into body.

Most people confuse spirituality with religion. We measure a person spiritually by whether they attend church or purport to do good deeds in accordance with a godly life, rather than if they hold the radiance of "being." Religions around the world have tried to weed out the concept that an individual could be directly connected to God, without any intervening priest or ritual to translate for them.

Spiritual experiences are only authenticated if they happen within the confines of religion. This has created a huge amount of spiritual insecurity. We feel that our attempts at spiritual expression must be unworthy and too insignificant to be real. By its very nature, spiritual experience is personal and must be validated by one's own self. It cannot be judged by the outside world. Every flicker that comes into our consciousness, reminding us of divinity, is a precious beacon of light shining upon the truth of our Soul.

In the west, we seem to have disassociated nature as a part of our innate spiritual heritage, even though this has been the teaching of all indigenous peoples. We have forgotten the uplifting feeling that comes from contemplating the beauty of a flower, the power of a bird flying, or the sense that we are a part of nature's majesty. Nature opens us up to other worlds that still the mind and whisper the answers to our limited life.

Fortunately things are changing and we are beginning to realize that spiritual energies come from deep within us all, regardless of form or dogma: We actually inherit our predilection to them just as we inherit our temperament and hair color. Thus, we may review our lives and not perceive any special experience or knowing that makes spirituality an important conversation to us. Nevertheless, spiritual undercurrents shape our existence because the laws of karma direct us to the lessons that feed the Soul.

Each incarnation of the Soul carries its own blueprint as to the themes and teachings that will be predominant in the lifetime. It is a divine Soul that incarnates, not an accidental human. We cannot get into body without the Soul's mandate. Spiritual DNA, like our other DNA components, is threaded through our very Souls!

We hold our relatives in such confining perception that we rarely know anything about their inner or spiritual life. Because they don't speak of it, we presume it isn't of importance to them or us.

Grandparents, especially, are not asked about their inner wisdom, but rather delegated to mundane conversations about their health or insignificant chatter. We see them through the specific roles they play within the family blueprint and cannot conceive that they may actually be someone entirely different. They themselves may be unwilling to risk ridicule by exposing their deepest and most mysterious experiences.

In the last twenty years, as we have extended our life spans, there has begun a disheartening trend of placing the older generation in institutions that care for the elderly. The message is clearly that they have no further value within the family or even the societal context.

How tragic to waste the richness of a life, the wisdom of experience, because we cannot face the task of caring for a weakened body! It is understandable that Alzheimer's disease runs rampant in the twilight generation. We do not wish to hear their memories or thoughts, so they stop having them. The environmental causes of these degenerate diseases are not just the pollution of our food and world, but the poisons within our inner environments.

I find it interesting to witness the correlation between this reality in which the elderly are pushed out of their place of honor and the voices of young people who are crying out in angst that we will not give them room to BE in our world. Our educational process has been extended so long that many young people have given up feeling that they have anything to offer a world of such complexities. Life is too big and complicated for the seemingly powerless. The recourse is often drugs, which take our children out of an impersonal world into one where the magic still exists.

Because of the subtle and powerful channels that transmit energies through the ancestral Chi and spiritual DNA, younger generations are inheriting these thought forms of dismay and rejection from their grandparents, who would be horrified to think that they are transmitting such negativity. So many older people feel that they are just waiting out their lives with no purpose and nothing to give or do. Their predicament is echoed in the words of our young people: "I'm bored. I have nothing to do!"

Distance, death, and even time have no power over psychogenetic pathways. There is no way to protect the young from such covert energetics that are seeping into their psyches, except through conscious clearing of inheritance. It has never occurred to us that we might inherit—almost simultaneously—energies from others without our mental agreement, but we surely can.

Mostly grandparents seem too far removed to consider that we might be re-enacting their realities—especially since, at any distance, we know so little about what is going on inside them, even though we might think we do. It is striking to wonder that what we feel to be our emotions or conclusions, may actually be theirs!

Contemplate for a moment what you know about your grandparents. Are you aware of their thoughts or feelings? Are they pessimistic, cynical, or enthusiastic about life? It is unlikely that you know very much, as the older generations have not been given to discussing their inner worlds with either their children or anyone else. They have been ruled by a society with specific laws of behavior for men and women. Externally they may have never broken free, but internally we can only speculate about what they know or feel, especially in terms of deeper worlds.

Through their spiritual Chi, they have transmitted to you their inner conclusions and experiences. Your intuitive faculties can help you access the commonality between you and your grandparents. Even if you never knew them, or if they have already passed on, you could still be psychogenetically engaged in conversations with them which are influencing your life. It would be very helpful for you to know about their lives as if you were looking for the invisible threads that bind you to their experiences. They are there!

My grandmothers slipped out relatively early from this world, leaving my grandfathers to reflect on life's meaning for themselves. I always felt that each of my grandfathers was fairly explosive and angry, especially toward the end of their lives. As I view it from my present consciousness, I see that I too, was very restless and irritable. I tried to write, but I could not find words for what was inside me.

Not until this moment have I realized that there is a connection. I am not saying that it was their fault I was so volatile, or that it was only their negative energy. I am saying that I was expressing from a deep level the frustration we three felt about things inside us that sourced our spiritual experiences, which we could not release or transmute.

Both my grandfathers were brilliant philosophers for whom life did not unfold within their ideal. In the corresponding era, I was searching for answers to similar questions. No one in the family was aware of their thoughts and it was not until my paternal grandfather died that we found hoards of books on topics such as hypnotism and reincarnation hidden in his basement.

He was ahead of his time and as a professional, he was entrapped by all the hardware of cynical science. Who would have guessed? I realized that I had always known this about him; I just didn't know what it meant or how to join him on those levels. Remembering a few conversations we had, I realized that he had alluded to his awareness of other worlds more than once.

My maternal grandfather belonged to a secret society that actually uses such principles, but would never avow to them publicly. He was so narrow-minded and opinionated that no one would have thought him to hold such lofty ideas, yet I know he did. How I wish he had talked to me about what he knew. Here I am now, having walked a similar path. It could have been so comforting to me to know that someone in my family knew what I was looking for or at. He did talk to me about the great civilizations of Egypt and Syria. I just didn't realize what he was actually saying.

Your grandparents are the apexes of your family. As such, they hold a place of wisdom and choice. It is not so much the things they have done, but what they have learned from experiencing the cause and effect of those choices. Because they have lived through so many of life's challenges, they have much to offer, even though their experiences are from another time frame. You are the beneficiary of their experiences, and whatever they have worked through will temper what is passed down to your children.

I love to ask people about any fantastic abilities carried by their grandparents. Often they are extremely powerful and special. They may range from great artistic talent to skills of healing or invention. It is fascinating to observe that the unique gifts and skills a person brings to life are rarely utilized or even present in their children, but rather re-emerge in their grandchildren.

We could speculate that their children do not choose on some level to take them up because of their need to find something that is their own. This could be a very wise genetic process that insures that new and better adaptations will be introduced within each generation. However, it is often emotional resistance that simply wastes potential, for the benefit of a rebellious ego. When the quality surfaces again in the grandchild, the gestation of one cycle brings it into being through a new facet that can be expressed in accordance with present conditions.

Through an inherited talent, it may appear as if the ancestors are still orchestrating our lives, or bestowing their blessings on us. Members of a family may associate these predilections with their beloved and feel a certain closeness because of them. If a grandchild shows an aptitude for that talent, he will be encouraged to develop it and the family will imprint him with the legacy of the grandparent.

A most intriguing phenomenon is the case of a

grandchild who has never seen its grandparent and yet makes the same gestures and has identical tastes and personality traits from an early age. Many families have whispered that they felt sure the grandparent has been reborn as the grandchild.

From a spiritual perspective, the concept that a parent could reincarnate as his or her own grandchild is perfectly plausible. The parent Soul would be experiencing certain themes for growth purposes and those themes might not have been sufficiently completed at the end of the lifetime. Oftentimes when someone is dying, people make emotional vows such as, "I won't let you go!"

The one who dies passes into the astral and is basically held there until that vow can be acted upon. When there is an opportunity to re-enter, their spirit will be magnetized to their own grown children. Through those spiritual contracts, the parent could be born as a child of their child, so that what has begun can continue.

If a family sees a child in this light, they will reinforce any innate similarities and strengthen the psychogenetic pathways that reconstruct a new version of the grandparent. Perhaps it does not even matter whether it is an actual reincarnation of the grandparent or not. The most relevant point is how the relationships are repeated or evolved. That we live on through our descendents–is certain!

Beyond the inheritance of our grandparents lies a path back through the ancestral Chi into the wave upon wave of energetic material that has sourced a thousand lives. From them all has been gathered an almost infinite array of qualities and gifts that we can call upon to enrich our lives.

To enter, in some way, into those lives is a breathtaking adventure. To imagine that each of those lives could be a stream of incarnations of the Soul that are connected to you may be unfathomable and yet somehow anticipated. Your grandparents are a window into an older world, another life experience. Can you imagine lifetimes you have spent with them that have brought you together in this one? They almost always have a quality of spiritual energy and the themes that come up are usually of a more expanded nature, like profound teachings being directed to you. Sometimes, the theme is about the fear of living spirituality in life. It is time to release this point of separation from your DNA.

At The Light Institute we follow the clearing of the grandparents with the Inner Grandparent essence sessions. Essence sessions allow us to access the universal energies that we hold as the head of the family or tribe; the wisdom keepers. Decisions that affect large issues in terms of the protection or direction of a society are the illuminating points of these incarnations.

The opportunity to experience this kind of energy can change you forever. Without having to wait for such wisdom in this life, the ability to wield knowing is a gift that comes to you through the power your Soul has accrued in other sojourns into body. By revisiting them, you find yourself expressing a higher essence, which has always been within you.

Perhaps you had the good fortune to meet or hear about your great grandparents. They would carry an even more purified spiritual Chi. The equation "less is more" is descriptive of the subtle currents that come from three generations away. The distance reduces the conversation of how things should be done and accents a deeper level of truth.

It is not so much the mundane aspects of what our grandparents or great grandparents have done in their lives, but the energy they hold for us. Because consciousness never dies, they continue to learn, sans body, and pass the teachings on to us through psychogenetic inheritance.

Our grandparents become the windows to the sky through which we can be inspired to search for a deeper truth. Even if we have never thought much about them, they anchor us into a bigger world, a more expanded sense of who we are. It is magnificent to touch them through their essence of spiritual DNA.

Chapter 8

SEXUAL GENETICS

It is much easier to look around our family constellation for the telltale signs of physical or even emotional inheritance than it is to ponder our sexual genetics. Of course we know that we have inherited sexual conversations from our parents and siblings, as well as our cousins and friends. We just have never thought about what it means.

It is so delightful in sessions when a person is observing their own conception and experiencing their parents' feelings during the sexual act. Very often they are totally surprised by the passionate, joyous and blissful juice that flows between their parents as they bring in another life. Conception is wondrous magic and its cosmic intention is imprinted in all of our cells!

Unfortunately, adults cover up the signs of their passion and, as children, we experience the denial and avoidance of such conversations, which leaves us uncertain that we are a part of something so very good.

To even imagine the sexual inheritance we might have received from our parents could be unsettling to many of us. In generations past, there was so much sexual guilt and lack of communication that we had to learn about sex on our own and only wondered the unthinkable about our parents!

Even when we didn't know how to phrase the questions that would tell us about our sexual genetics, we still wondered. Were they passionate? Did my mother have orgasms? Was my father a good lover? Do they still "do it?" The probable answers reflect on us and what we think is possible within our own experiences. We worry about what we have inherited from them. We can see that our bodies are copies of theirs and so we watch to see if it means that we will be great lovers and coveted partners, ourselves.

It is much easier to enumerate what we don't want to inherit than what would be wonderful to have received from our parents, because we know so very little about their experiences in our outer awareness. What went on behind closed doors (which could have been the hottest romance) may appear to us, the observers, to be the most distant of acquaintances.

Sexual genetics form the structure of a thousand questions we ask about ourselves: how we feel about our bodies, whether we feel worthy of love, do we really know how to make love? All these questions revolve around instinctual, primordial energies that move inside us and have been passed down by a hoard of others who suffered the same pangs of self-consciousness as do we.

Virtually all of our core themes stem from sexual genetics. We cannot separate the theme of sexuality from those of relationship, power, rejection, or even money. We wield our sexual energies in all these arenas as the great neutralizer.

Ask yourself how your sexuality has attracted or repelled relationships. How have you used it—as a weapon or a divine gift? How do your sexual activities make you feel about yourself?

The answers to these questions begin in the early part of our lives. We are told what we cannot do with our "private" bodies before we are even five years old. Hardly a child has not learned that touching his or others' sexual organs displeases parents and upsets adults. We have discovered that our pleasurable bodies are a threat to almost everyone, yet we are aware that our parents have some secret about their bodies that they only share with each other. We want to be a part of it, too.

Perhaps Freud was close to the mark when he brought our attention to the Oedipus stage of development, wherein boys want to marry their mothers and girls their fathers. The attraction is about experiencing whatever that secret is between them. We want to belong to our parent the way they belong to each other. The sexual energies are more diffused at that stage of childhood, but they are definitely there.

When we grow up, we choose sexual partners who most "feel" or look like our parents. Perhaps we would not admit that there could be any connection, but somehow there is an aspect of playing out the childhood fantasies with someone who fits the role.

I have listened to the sorrows of women who

relate how fantastic their sex life was until they married their lover. After marriage, the man seemed to lose the sexual connection, as if he were a totally different person. Sometimes the reason for this is a subconscious crossover between his feelings towards his mother and those he holds for his wife.

Usually he has chosen a woman who reminds him of his mother, either by looks, character attributes or behavior. The formality of marriage may have triggered a deep unconscious projection in which he feels he is doing something with his mother. It's enough to fester hesitation, which is disastrous to joyous lovemaking. For some men this continues even more strongly with the birth of children.

Seeing his wife somehow as his mother, he could find the honor and sacredness, but the passion had disappeared almost as soon as she placed the wedding band on his finger. In some cultures, mothers are left to raise their children and the father begins to wander about, feeling somehow displaced sexually by the unconscious parental lust. Either the woman is the sacred mother or the "woman of the night."

The famous Austrian psychologist Karl Jung said, "Ask a man how he feels about his mother and he will tell you how he treats his wife!" We carry on with our partners a replication of what we have witnessed in our parents during our childhood. The man will look for a woman who will play the role of his mother so that the emotional energetics can stay the same. Women have an equally repetitive scenario, in which they search for a man who will continue the play they had with their father.

The powerful illusions in this relationship script lie

in the emotional pattern of *projection*, in which we see in others what belongs to ourselves, or what fits our emotional repertoire. We then project upon our lovers what we have learned from our parents.

When there is a special relationship between parents and their children of the opposite sex, there is sometimes an entanglement of unspoken vows and contracts. Both parties may promise consciously or unconsciously, to be loyal to each other forever. Long forgotten, the intention lodges itself into the fabric of emotional DNA and cannot be extricated because there remain no conscious reference points.

Thus, the son dutifully can find no one who measures up to his mother and the daughter feels a vague sense of danger and holding back when she makes love with her husband, but has no idea why she is so reserved. The misplaced loyalties are coming from long-standing agreements made in forgotten times, that replay themselves early in this present life.

If your parents hid all hints of sexual exchanges from you or taught you that sex was dirty, against God, or dangerous, you are very likely to re-enact those conclusions through the way you engage sexually, even if you disavow the connection between you and your parents. These imprints become your inherited sexual genetics.

Take a moment now to contemplate the messages you have received from your parents about sex. What do you think are the thought forms your parents have in regard to sex? It would be very worthwhile to write down all the thought forms you have ever heard so that you can sort through and see which ones have stuck to you on a deep level.

For example, in your outer consciousness, you know that "sex is not dirty," but you may hold back some part of your being when you make love because the residuc of that thought form is affecting your ability to tune in completely to your body.

When you find a thought form that is adversely affecting you, you can go into your DNA and dissolve it.

Take several deep breaths to put yourself in a meditative state.

Ask your body to show you where that thought form is lodged in your DNA. (Accept whatever image comes into your consciousness. You might see a spot or a clump on a rope or a sparkler. There is no right answer, as each person will perceive it in the way their body shows it to them. Every time you go into your DNA you will see it differently, just as the 50,000 gene coordinates align, depending on their purpose.)

When you have found it on the DNA strand, ask your Higher Self to select the perfect frequency of white light to dissolve it.

Imagine that your Higher Self is lasering that brilliant light into the DNA and dissolving the thought form.

Take another deep breath and feel the lightness in your body. You are becoming FREE!

Every time you consciously remove a negative thought form from your DNA, you are changing the future for us all.

Sexual genetics give us a wondrous view of the connection between physical, emotional and spiritual DNA. As embodiment only occurs through spiritual intention, the procreative urge whispers to the primordial body and brings about emotional expansion, which opens the threshold through the power of feeling. The Divine comes into form through sexual energy, and therefore sexual energy is the closest energy to spirit.

How could the Divine manifest form without bliss and cosmic celebration? If we are to discover the true energies of sexual genetics, we must extend our consciousness to the teachings that come from Spirit. They will liberate us from the misinterpretations that have stifled our experiences up until now.

It is only logical to accept that if sexuality is the vehicle of transforming divinity into body, it could not be "bad" or displease the Creator, who designed it that way. Yet, the guilt and shame persist—not because there is any truth to sexual impurity, but because they have been instilled into our genetic patterns by repetitious influences. They were taught to us by those who were branded by the fears and threats levied at them from others who, themselves, were terrorized about the danger and shame of sexual expression.

All religions are united in their positionality that sexual energy—at the very least—dissipates devotion to God. They hold that sexual activities distract us from devoutly focusing on the Divine. On some level, that observation has translated down to us that sexuality is in contradiction to God. Then who or what gave us the fantastic rush of sexual pleasure? ...Why?

I would never be convinced that sexual joy is only to ensure procreation. It is there before we are able to physically conceive and long after the possibility that babies could be the result of such sumptuous enjoyment. Sexuality in all its glory is our birthright and the sweetest gift of our body, with which to savor life!

Many of the negative thought forms we hold about our sexuality come from cultural as well as familial constructs. Like a colossal conspiracy, churches, schools, and societies at large have barraged us with messages that sexual expression spells automatic separation from our own goodness. To the other extreme, movies and the media send us the almost irrepressible confirmation that sex appeal is the only way to success.

In truth, almost everyone we know influences our sexual genetics. As we perceive the conclusions of others, we add each one to the tally that builds our own conclusions. From sessions on sexual genetics, I have discovered that people who consciously feel they are in balance with their sexual energy are still carrying a tremendous amount of twisted emotions and "unspeakables" that they have acquired from the world around them. They have dispelled them from their minds, but the energies are still lurking around beneath the surface, causing discomfort and disease.

When these experiences become part of our way of being, they enter our DNA as permanent aspects of who we are. The DNA is the essence of our life force. It holds the potential of what we have gathered from the thought forms and experiences of others and locks it into an incredible, almost infinite pool of possibility.

In the same way that we imprint emotions from people around us, we have no protection from taking on the sexual encoding of those with whom we share sexual energy. The thrust of sexual frequencies is so strong that we also take in the imprints of any other lovers they may have. A "one night stand" penetrates and mixes the auric fields of two people for about 48 hours. The long-term residues last almost nine months!

If you have a lover from another culture, you will take in many aspects of their realities through the very act of making love. Generations of sexual attitudes and societal codes will begin to mix together. As that energy flows through your auric field, you will seem more familiar to them and by the mechanism of horizontal inheritance, you will each seed each other.

Around the world, sex is seen as threatening to God and home. Even in our "free" societies today, there is an undercurrent of guilt and shame that has not been dislodged by sexual experimentation or rebellion. Nor will it be, until we clear it from our DNA!

The fear of sexuality may be stuck on your physical DNA helix or the negative associations of guilt may be linked to your emotional DNA. The thought forms that sexuality separates you from the Divine may be encoded into your spiritual DNA.

Ask your body to show you where on your DNA you are holding sexual guilt or shame.

Laser it with white light until it is vaporized.

Sexual encoding is so intertwined with our sense of self that our Emotional Body uses sexuality as a way to hold sway over who we are. It becomes entrapped in defending or protecting an image it has created to enhance its viability. The façade separates the Emotional Body from its spiritual awareness.

This is a crossroad of emotional sexuality. It can either become a thrilling cosmic experience without boundaries, or succumb to a thousand fears. If the Emotional Body uses sexuality as a weapon of personal power or protection, we cannot discover the ecstasy of higher sexual frequency.

The jubilant states of saints and the bliss of new lovers both come from the delicious sexual currents that rush out of control within our bodies and lift us up into heightened awareness. Through these, we become sensitized to the rarified octaves of true sexual potential. The world does, indeed, look more beautiful from there!

Sexual energy is such an exquisite power. What shall we do with it? Within the veils of consciousness, a thousand times a thousand healers and teachers, artists and priests have transmuted or transcended the outer manifestation of sexual energies and brought that force up into a higher octave of expression.

Could you imagine such rapture as to be levitated off the ground by the force of sexual/spiritual energy? We have heard of it happening more to people whose lives are filled with Spirit, because they do not "spend" their sexual energy in the physical realm. If it happens to them, it can happen to us!

The play of our sexual/spiritual energies becomes the divine "Shakti Body" that brings together the primordial sexual urges with the light frequencies of the spiritual body. The Shakti Body provides the nectar or juice energy that feeds the force of kundalini.

Our sexual source stems from the famous kundalini, the coiled snake spoken of in eastern traditions, seated at the bottom of the sacrum. As puberty triggers the master endocrine glands to activate sexual maturity, the energy rouses itself to rise up and return home to the pineal gland where life purpose and spiritual vision are held. The kundalini has an outer layer that becomes our sexual current and is spent in the act of making love. It also has an inner layer of pure energy that returns through its pathway in the spinal cord to reunite with the "third eye."

I have had two experiences of spontaneous levitation involving my sexual/spiritual energy in this lifetime. The first one was after the third time that I received *shaktipat* from Muktananda, a famous Indian guru. Almost immediately after being bopped on the head with his peacock feather, I flew into a full lotus position and rose up about a foot off the floor. I had been sitting in deep meditation and was already in an exquisite state of bliss.

The humor of it was that the levitation felt absolutely normal to me, as if weightlessness were my true expression. On the other hand, the experience of a perfect lotus position with the erect spine and the legs folded into my groin created a shock wave through my body. It caused a force of light that streamed up through me and out my crown chakra. I felt like a living candle. This is a perfect example of the divine sexual potential expressed through its purest energy. Through the power of this new frequency, I awakened to a spiritual inheritance that changed my life, although I didn't understand what it meant at the time.

I have told you about the second levitation experience in my book, *Ecstasy is a New Frequency*, which was a direct result of sexual energies. I was lying out under the Arizona stars on Christmas Eve with my five children. My last thoughts before sleep had been an aching sadness that I would never use my sexual body again. Sometime later that night I was given a prophetic answer that eclipsed all my other sexual repertoire. Lifted from sleep and from the ground, I was radiated by a trillion pinpricks that caused my body to undulate in an electrical pulse. It was Cosmic Orgasm!

It felt as if all the stars were kissing me, holding me in their embrace—not the gratification of being loved by a person, but a love with intensity beyond anything two beings could create. The sensations were not connected to a part of my body, but my body became a cosmic coalescence—and then, ecstatic ripples without an end...

The levitation felt so much like my natural state that I hardly noticed it until I experienced the caressing, gentle floating back to the ground. It was but an aspect of the whole experience. I cannot guess as to whether I was there in those cosmic frequencies for seconds or many minutes. It seemed quite a long time and towards the end I realized that I was a good distance up in the air. Part of my consciousness was aware that the sides of my purple sleeping bag were swaying in a breezy, dancing motion.

Twice, while people were doing sessions on sexuality in my healing room, I have witnessed levitation. Both times they were experiencing profound states of bliss after transforming their sexual energies into spiritual frequencies. It was not that they were doing this mentally or visualizing it; they were directly experiencing a repertoire shown to them by their Higher Self—beautiful lifetimes in which their sexual energy was aligned with their spiritual energy. They were actually encompassed in those frequencies. Each time, they simply floated up in their horizontal position, as a result of the absolute quickening of their energies.

We have collective memories of saints who levitated, who became so ecstatic they couldn't stay on the ground—cosmic orgasm, magnificent points of enlightenment that allow us to be part of the cosmos. We have inherited all of these. They happened when people made love and they happened when people connected to the higher energy. From them all, we have inherited a Shakti body that can give us the same and even new levels of sexual expression.

It is only now that I am realizing how we could invigorate our own encoding with these kinds of experiences. By touching them with our consciousness, we might open the windows to infinite memory banks that would help us all make the leap up to such states of rapture. Through them, we could transform sexuality into something that embraces our divine source.

Perhaps you are wondering how you could access

these energies. You may think that no one you know has ever had them. I promise you that almost everyone you know has had at least one exquisite sexual experience in their life, even you!

You may not have levitated or felt that you were having a divine experience. It might have been simply a thrilling electrical current or a brief moment of feeling merged with someone. If you focus on the energy, not the memory, you can repeat it again and again. At some point it will imprint on your DNA and become a focal point that magnetizes corresponding energies to you. You do have good and loving sexual imprints; you need only to ask your body to show them to you.

It is time to erase our mediocre sexual residues in which we were waiting for someone else to make us ecstatic. It is time to fearlessly be present in all energetic exchanges. If a saint or enlightened being experienced sexual ecstasy, they have left their tracks for us to follow. The most important clue is that of consciousness. If you imagine that you have received their blessing in the form of genetic encoding, you are already open to having wondrous experiences yourself. It is simply a matter of choosing them.

Try this exercise in consciousness to view more of your own sexual genetic repertoire:

Ask your Higher Self to show you a powerful sexual attribute you have inherited. (It might be the gift of orgasm, the ease of merging, the lightness of heart, or ecstatic rapture.)

Ask to be shown when you have used it in this lifetime.

When the memory comes to you, allow yourself to become immersed in its energy.

Ask your body where it is holding this memory.

Bring your consciousness to that place in your body and free the energy so that you feel it flooding through your body, imprinting it in all your cells. Take a deep breath and let the memory itself go, so that you can create new reference points now.

To contemplate that we have inherited sexual attributes from beyond our family constellation, from beyond the limitations we perceive as human, is awe-inspiring. We know that we are part of something vastly bigger than our small world, but we have not yet grasped its potential. It is almost unfathomable to consider that sexual essence coming from the divine universe flows through us and that we are encoded with its pure energy. Just to wonder what it would be like to touch it, to be embraced by it, could bring us infinite joy!

Here is a higher octave exercise of sexual genetics:

Ask your Higher Self to show you the highest sexual encoding you can become aware of at this time.

You will probably experience it energetically. (Typically, people feel dizzy, flooded with joy, heart opening, sensations of floating, flying, swirling.)

Let the energies move through your body and then see if the sensations carry you into any images or experiences. (You could see yourself performing a miracle, loving or healing, singing, being an angel, or feeling the Divine moving in you.)

Ask your Higher Self to show you where this energy is located in your DNA.

Ask your Higher Self to laser the highest frequency of white light into this point to amplify and awaken it within all your cells now!

All experiences of any human being are consigned to our pool of genetic potential. We can learn to access them within our DNA and bring them to life through our consciousness. By threading back our DNA to the memory banks of all ecstatic sexual experiences, we can establish a record library that would show us how to translate sexual currents into ecstasy, rapture and bliss. Through these states, we could perform all manner of miracles; heal ourselves, touch our oneness, transcend time and space!

Chapter 9

CULTURAL INHERITANCE:

THE POWER OF THE TRIBE

The word "culture" bestows a sense of time-honored tradition that is comforting to our need for something refined and uplifting within our view of life. Many of us today are hoodwinked into thinking that culture is about styles of living—music, literature or other pastimes that place us into secure categories or niches of prestige within our societies. We have long since cut the strings of external control over our choices and think ourselves free from outer surveillance. We feel this way because cultural inheritance is so intrinsic to us that we do not recognize ourselves as the product of societal conditioning.

However, in the same way that our familial experiences shape our sense of self, our cultural inheritance also defines our self-image and worth. It is an environmental permeability that sponges up the soup of collective consciousness and pours it through the sieve of our individual personalities. Our feelings about art, beauty, sexuality and community are all the filtered precepts of our cultures.

The most relevant aspect of cultural inheritance is its scope of influence over our sense of self. Its propensity to pre-package an assembly line of compliant people who see themselves as "ordinary" is monumental. The very term *ordinary* conveys a kind of security in mediocrity that supports conventionality rather than individuality. It is the cultural stamp of approval that boxes in our sense of adventure, the thrill of inventing and re-inventing ourselves.

Cultural inheritance includes a thousand times a thousand innuendoes, gestures and collective thought forms that have been agreed upon by the members of our culture, back to the very beginnings of our family tree. They are the prerequisites for acceptance within the arms of its exclusive embrace and the encoding is so insidious that we simply do not recognize it. While on the one hand, we think they are the infallible truths of our lives, on the other, many of those thought forms are individually stifling and painful to those who struggle underneath the weight of such a prefabricated existence.

Thought forms that once held together the fabric of the community are today binding and even counteractive to the recognition that we are, each and all, a unique holographic conglomerate of the whole. In the past, societies around the globe expressed themselves through tightly defined patterns of behavior that precisely delineated the appropriate response to life, according to one's gender and stature within that society.

Why not explore the cultural thought forms that

are shaping you now?

Ask your Higher Self to show you the cultural thought form that is most inhibiting your growth now. It may be about sexuality or visibility or fear of others.

When you become aware of it, ask to be shown where it is encoded on your DNA.

Allow yourself to perceive it and then laser a brilliant white light into that spot until it has been dissolved. Take a deep breath and feel the lightness that comes of freeing yourself from its grip.

Do you think you could describe yourself from a cultural perspective? Have you ever wondered why you were born into a German or American or Latin culture? What were the lessons you have chosen to learn from your cultural genetics? Cultural themes are identical to all themes of humanity. Different cultures may offer a specific focus such as rules for community, sexuality, spirituality, justice, freedom and materiality. The divergent perspectives always serve the lessons we require on spiritual levels.

One of the biggest conversations in cultures around the world is the new role of women. We are emerging from a dark night as invisible beings, into a formidable group of wayshowers. The transition has been wrought with emotional and somewhat extreme measures to cast ourselves out into a world not really willing to receive our gifts.

Women are seeking a place of equality that frees us to contribute something from deep inside us. We know how to make things happen and we often see the solutions, but we keep waiting for permission, as if we were children.

The problem lies in our need to win approval. We either try too hard to be the "good little girl" or we rebel and use the old masculine (yang) tactics that recreate the separation issues of the past. There are new choices, but we must rid ourselves of the inbred cultural mandates that cloud our purpose and our hearts.

Some of the pain and struggle comes from the imprints we have carried since childhood. Clearing cultural inheritance in sessions has shown me that almost all women have the theme that women are less valuable than men. It might be the feeling that the father wanted a boy, or that boys get to do more.

In every culture, there are very real differences between how parents and the society at large treat boys and girls. These are nothing more than habitual residues perpetrated through the myths passed down, generation to generation. Ultimately, we women will have to change the way we relate to our children in order to circumvent the predictable outcome of how our children's children are raised to view themselves.

Let's clear our cultural thought forms about women now so that we can contribute to instilling a new perspective into the future.

Take several deep breaths into your brain and relax. Ask your Higher Self to show you any inherited thought forms about being a woman or about women, that need to be cleared now.

Ask your Higher Self to locate the exact point on your DNA that holds the thought form. Laser the most brilliant white light into that place and vaporize it.

There is a "ripple out" effect in terms of collective consciousness. By releasing these old thought patterns, others will change theirs by sensing that there is another way to be. We have felt our lives and relationships dictated by our cultures for too long. It is time to realize that it falls on us to change our cultures, not the other way around.

Consider the pressure felt by first-time parents that their child be healthy, intelligent and a model example of their culture. All of the subtle rules of parenting bear down on them to make sure that societal belief systems are passed to the children so as to continue the status quo.

In their insecurity, young parents allow others to dictate how they raise their children. In turn, the first-born suffer the brunt of those mandates and often feel like adults who don't know what it is to just play, without concerning themselves with the needs or mandates of others. The weight of cultural approval often makes first-born children strive to be more successful than other siblings within the family, but more conflicted as well.

Cultural indoctrination begins early in life. To the open psyches of our children, a look or a word is more than enough to signal our approval or disapproval of others. Rather than engendering fear, it would behoove us to carefully teach our children the qualities of openness and adventure in communion with all beings—collectively and individually.

We need a catapultic leap of consciousness to glimpse possible new cultural models. So far, we have thought ourselves brave because we are trying out each other's parts in the play. It is a very horizontal perspective that urgently needs to give way to a vertical shift into a whole new universe of being.

Today it is necessary for each of us to play at least several roles to attain a successful sense of self. We can no longer be just the mother or the teacher or the husband because we, ourselves, are growing so rapidly that it is impossible to be contained in one channel of relationship. How can we find nourishment or inspiration within the confines of singular expression? We are *not* just women or men, mothers or employers. Where once we could separate ourselves into different realities, such as work and home, now we are reaching beyond the walls that entrapped us and are seeking the freedom of *spontaneous selfhood.*

Though societies appear to change rapidly, the rickets of porous cultural bone go through a prolonged process of disintegration. Emotional associations and conclusions of our ancestors seep into our very being and dictate the tides of our reactions. Encrusted emotion is the bonding material that holds us together and we cannot possibly tell it apart from that which we feel to be our own.

Even when we think we have forgotten our cultural origins, the residues and imprints whisper to our every choice. The habits of life are so structured by emotional programming that our "cultural bodies" are in a state of rigor mortis and we have not noticed that they were ailing!

The recognition that emotions are the insidious weapons of cultural battlefields is very profound. We are emotional about our flags, passionate about our sports teams and violent about our religions. All religious expression that binds to cultural peculiarities is based in humanity's profound need to live a life that responds to the mysteries of God. The essence of all cultures is the experience and incorporation of the mystical and the Divine into daily life.

From the very beginning of tribal culture, art and ceremonies fulfilled man's deepest needs to communicate about and create channels of relationship between themselves and their gods. As tribes began to interface with one another, they forced each other to include or accept their gods. Ultimately, cultures have used religion as a pretext for whether they could relate or not. "My God against your God" has been echoed in almost every battlefield. The power of the tribe holds sway over the passion provoked when its autonomy is threatened. Lethargic masses have been whipped into a fury by the indignation of differences, so much so as to throw life into the fray.

In today's world, religions are beginning to transcend cultures. The major religions have participants from all cultures who adapt, to some extent, the expression of the religion to their local ways. Religion is becoming a global conversation and it is crucial that it be reconfigured back into a more spiritual energy that permits each person to experience the mystery of the Divine in a natural and direct way.

As we make these broad sweeps over time and space, we realize that our cultural repertoire spans all of history. Tribal systems are as much a part of our cultural inheritance as present day societies. The primordial residues of tribal existence have instilled certain codes of ethics and life patterns in us that could both facilitate our coming together and destroy us through the illusion that our differences are too dangerous to survive.

We are as tribal about our friends and families as we were thousands of years ago. Within the psychogenetic web of humans are all the activities in which we have engaged from the beginning of human gatherings. Time holds no power over genetic inheritance and only a mutation that alters the pattern has any effect on the surface of the form. This illumination is both deeply disturbing and promising.

Tribal life has been filled with monotones of neutralized expression wherein the focus was on repetitions and singular roles. The primordial and real fears of “tribe against tribe” are resurfacing today in deadly duels of right to life. Ancient vendettas of ethnic adversity are causing terrible bloodbaths to occur. They are part of a myriad of imprints that are blocking further evolution and thus are being triggered into mindless activity. These inherited hatreds have become amplified so that we can realize their innate destructiveness and dissolve them from our DNA.

We must stop them! The only way to see past

the polarities between groups is to expand our consciousness beyond the limited perspectives they hold. Imagine the poisons we carry from such putrefying encounters. Whether we are infected with such lunacy by virtue of blood relatives or by having fought the battles in other lives, we are indebted to future generations to protect them from the fatal moves of humans that endanger our planet and our species.

Close your eyes for a moment and imagine a person from another culture. It could be an Aborigine or a Greek, an Asian or a Dane. Simply allow yourself to see who comes into view.

Think about what you know of this person's culture. Do you admire them, find them attractive, or fear them? Don't try to use your intellect. Let thought forms and images surface in your mind's eye. Can you fantasize about their life? Who do they hate or fear? Whatever was theirs is yours, too!

Cultures of the past and even those around us seem so far away from what is important to us. Entrenched in our own realities, we may never give them a thought. It is a new concept that they could be connected to our destiny or the way we live our lives, but they are. We are in tandem with all human cultures and will always be interdependent. What they felt or do now creates a sounding board from which we hear and see our own lives. We may not feel that we care, but we must, because the flicker of their experiences has become the flame of our potential.

This is where cultures have become stuck. The interlocking emotional and spiritual DNA is communicating within the gene pool and dictates choices that hold individuals together, based on old cultural models. The mandates worked well within the isolation of tribes, but now that we are touching each other on a global level, those systems need to be cleared so as not to become detrimental to a higher fusion of humanity that must now outdistance war and harm.

Thought forms shared by whole cultures create entire lineages of similarity, which are passed down and reinforced by the collective perspective. *Learned inheritance* becomes inbred into the genetic fabric!

Contemplate that you and I, through inheritance and embodiment, have taken on imprints of vengeance and hatred. We are entrained with the "them or us" conclusions of the past. It may be almost impossible to relate to violent feelings inside us and yet we are witnessing the reflection of such things all around us. Through words and manipulation, we have many subtle ways of enacting our depreciation of others.

If we have no awareness of these negative residues lurking beneath our civilized exteriors, or view them as belonging only to others, we will project them onto almost anyone who seems slightly different from us and thus re-enforce the destructive flaw of defensive separation. It is crucial for us to ferret them out of our psychogenetic fabric so that we learn to embrace the wonder of many cultures without fear!

Ask yourself, truthfully, what fears you hold regarding other cultures or societies. From whom have you inherited them? They may have been appropriate for your grandparents or great grandparents, but for you they are a burden of unsubstantiated truth that separates you from the adventure of life.

The cosmic giggle is that we are linked to all races and cultures through our psychogenetic inheritance. We have reference points through our ancestors, our incarnations of the Soul and our earthly gene pool. We hold the proud heritage of one and fear of another because of forgotten memories that course through the stellar spaces of our helical essence.

We don't know why we love everything about Africans, Indians, Egyptians or Greeks. There is a resonance stirring within us that comes from some place so deep inside that it has no face, but the flickers of smell or image arouse us and cause us to reach out. We don't know why another culture brings us feelings of joy or sorrow, but the unsolicited emotions are not accidental; they exist on the chains of historical DNA.

Our feelings are the subjective conclusions of bodily experiences, not immovable truths. How foolish to presume justification for hatred or prejudice, when uncovering our real relationship would show only the warp of time. Our cultural posturing is utterly absurd in the light of genetic relativity.

Humanity has always embraced extremes as an expression of society. We applaud war as an excuse for unbending differences. Cultures, in effect, are a larger mirror of gang psychology, which insists that one inflict pain on another to show loyalty to the brotherhood. In the face of an uncaring world, we go to further extremes to belong—anywhere, to anything that relieves us of the pain connected to aloneness. How willing we are to "sell our Souls" in the exchange!

Consider the torturous rites of passage that came before acceptance as an adult in earlier times and then our present "non passage" with the emptiness and loss of belonging that plagues the young of today. Most people would prefer the drama and the pain that end in clarification of status to the loneliness and uncertainties that precede adulthood today.

Perhaps that is why we actually seek out other groups and cultures that challenge us to "prove" our commitment and loyalty to their rituals of life. Groups often fall prey to the standard of the lowest common denominator and engage in collective acts that bring karma to bear on each individual. Why do we do it? It is certain that a trait instilled by cultural inheritance will not be erased merely by viewing it with our minds. We must venture out of the fold where we can discover what is ours because we have made it so, by our own choice—and what we have inherited that we wish to keep as ours. There will not be an era of peace until all that is past has been cleared of unusable particles and reconstructed within an enlightened helix.

In the end, though we feel the mandates of our cultures, we ourselves are the instigators of Cultural Evolution and the designers of cultural constructs for generations to come. Cultures do not change rapidly; therefore, we may not ever see our effect on the future, but it is there. We are an emerging species whose individual revelations impinge on the rest of our collective family of humanity. Everything we do has ramifications that extend beyond ourselves into our families and cultures.

The restrictions of societal roles have grown heavy for those whose evolutional tiger is chasing after the aged meat of sedentary existence. We are past the thrill of being stalked and long for flight, not from our predators, but towards something free and light.

Chapter 10

GLOBAL INHERITANCE:

THE FATE OF HUMANITY

I was sitting in a group of eight young people from around the world who were preparing to leave on a journey to the North Pole as part of a project called Pole to Pole 2000. It was created by my friend Martyn Williams, who is a world renowned explorer and expedition leader to both the North and South Poles. He envisioned gathering a group of strong young people who would travel from one pole to the other, on a mission to secure promises from as many groups and individuals as possible to protect Mother Earth and seek peace in all aspects of life.

Eight beings from seven countries came together to answer the quest. They were from the ages of 19 to 27 and each one seemed a bright and shiny star. As I watched their interaction, I could see the hope of the future. They were so willing to reach beyond the limits of language and culture to touch each other and bond into a cohesive unit, furthering the purpose of their mission.

Yet again and again, I observed how the mind entrapped them and blocked their capacity to grasp the intangible. I saw the spectrum of how familial and cultural programming lay the bases for mental rigidity in which the mind perceives anything new as a proof test or abstract concept and cannot allow the experience of it as evidence or possibility.

I am struck by how polarized we all are in terms of our minds and how this closes us off from the rapture of fully conscious experience. If we deny the mind's ability to envision and therefore experience directly through its sensory faculties, our world will remain a closed circuit that will eventually atrophy. Humans have been gifted with more than 70 senses with which to embrace life, yet we have diluted its thrill and richness by detouring all perception through the mind.

We have not served ourselves as humans by identifying truth through our limited intellectual, rational minds. The very definition of intellect is the capacity to problem-solve and discover the purpose or the workings of unexplained new phenomena placed before us. However, we have interpreted it to mean nothing more than the assimilation of only a certain stratum of information, rather than the correlation of data from almost infinite aspects, that impinge upon each other to create a complex reality which necessitates holographic thought to comprehend.

In so doing, we have begun to weed out the very quality of mind that is vital to our next evolutionary step. Without the "higher mind" that encompasses holographic awareness, we cannot possibly access the random chaos that synergistically precipitates cosmic manifestation. In short, we will trail behind too slowly to interact with the forces of change.

The Pole to Pole team entitled their presentations, "Challenge to Change." We think of change in terms of attitudes and awareness about our world, but they are actually external measures that are somewhat secondary in effect. We must change from the *inside out*, in the same way the body heals itself. The answers to our environmental and global issues are not just in new approaches, but in completely revamping our experience of being human!

As we expand our consciousness into holographic awareness, we will embrace new thought patternings that link perception to cause and effect, action and reaction, and clear the way to a new kind of intelligence. It will be inclusive of beingness and Soul as holding the answers to life's greatest secrets and challenges.

I feel that we are evolving a new brain that will serve these functions. The frontal lobe corresponds to emotional perception and it is through these channels that our intelligence and caring for others will develop new facets of human potential. The dolphins and whales use their frontal lobes as sonar connectors to perceive holographically. They have a special fluid in their frontal lobes that creates the resonance for such incredible transmissions.

We, too, have brains that are more than 85% water, and we can learn to utilize the conducting capacity of water to augment our perception. We know that the dehydration of the body and brain is the major cause of aging. By simply nourishing our brain with water so that it doesn't become dried out, we can avail it of the highest human potential for consciousness.

Imagine that you are washing your brain. As you envision holding it under a waterfall or in the sea, imagine that all negativity and toxicity are being drained out. Then image your brain brimming with fluids, juicy with the water that conducts brilliant illumination and intelligence.

Children are being born today with larger frontal lobes that carry the prototype for this new kind of human. I would say that they have "chosen" to transition us into this more highly evolved species of Homo sapiens. They are delicately wired with a new frequency of energy that must be carefully nurtured if the "mutation" is to take root.

Our educational systems have been based on interjecting as much information into students as possible. It may be that these next several generations will be attempting to interject information into us! They are being born into a world that will expand beyond our present scope of reality and we must find a way to help their minds cope with new information that we, ourselves, may not be able to grasp.

Our children desperately need to learn how to trust their own inner voice so that they are not overly swayed by the unbalanced influences of the media and their environment. The educational format will necessitate a complete change in purpose and application to meet the needs of the future. To that end, I founded The Nizhoni School for Global Consciousness, whose carefully designed curriculum is focused on awakening a profound sense of self and the deepest knowing of each individual, so that at any age the student feels a part of our complex world. It is a "Soul-centered" form of education that encompasses the conversation of "Who am I?" and "What do I have to give to my world?"

I teach a course called "Themes of Humanity," in which we explore global themes that are part of our collective consciousness. Love and power, fear and hate, death and God are all subjects that shape our cultures and define our world. We begin to discover that our inner conversations are echoed out across the seas and touch all other humans. Through this exploration we can actually experience how it is that we are one global family.

We might ask ourselves which "Themes of Humanity" most affect our global relationships. I would begin with Religion, as it has done more to separate us and create wave upon wave of hateful vengeance than any other theme. The pretext of fighting for our God and destroying others because they have their own God has embittered whole cultures against each other. To use God to justify the plundering and stealing of others is the gravest of human travesties. All religions must accept responsibility for their part in luring their worshippers into this despicable fray. The epic of such folly is over. We individuals must stand our ground, both to governments and churches that would draw us into wars that are the antithesis of all divine teaching.

If we could release the mental entrapment of religious programming, we could embrace all forms of worship chosen by our cousins and friends. Our collective experiences of God or the Divine force would only enhance and enrich embodiment, rather than provide any kind of wedge between ourselves as humans.

The spiritual focus on the Divine is the truest and most powerful force to bring humanity together, not to alienate us one from another.

Our sense of fear and separation stem from learned belief systems, not from actual spiritual experience. Back in the shrouded beginning of religious expression we were imprinted with the concept of the "angry God" who had to be dissuaded from wrathful vengeance upon mankind for their imperfection. This resulted in the ritualistic practice of sacrifice to appease God.

I can truly imagine how easy it might have been for a priesthood that overstepped the boundaries of directing others' lives to fall into the trap of blaming their behavior for any mishap that befell the group. How convenient to say that God is angry and we must pay by suffering. This concept has buried fear and punishment deeply within the human psyche and into the spiritual DNA.

It is not God who is angry with us; it is nothing more than an effective manipulation that allows the few to control the many. Virtually every type of authority has used this technique to harness in its constituents—from priests to police to parents. The goodness in us needs to be allowed, without coercion. The era of martyrs, sacrifice, and punishment is over. It is time for us to realize that we can give a holy gift without taking a life or inflicting pain upon others or ourselves. We can instead celebrate the sacredness of life as divine expression.

Healing this global rift must begin with individuals who are willing to risk opposing the subtle coercion of society, in order to teach their children a kinder way.

Ask your Higher Self to show you the exact point on your Spiritual DNA where the thought form that God is an angry God is entwined.

Ask your Higher Self to show you the brightest white light frequency that will remove the thought form.

Laser that light into the point on your DNA and erase it completely.

Take a deep breath and feel the divine love that can touch you once you are not locked into the fear and punishment of the angry God myth!

Thank you, my friend, for doing this. It will make a difference to all our futures.

The incredible spewing of human hatred, separation and violence on our planet right now is the result of an influx of higher energy that is affecting our DNA. The increased radiation, the sunspots, and the synchronicity of evolution are disrupting the sediment of ancestral imprints. The prophets of every culture have locked into this time frame and spoken of "brother against brother, tribe against tribe." It need not be that way!

Like a powerful centrifugal force, the altercation of frequencies is spinning the residues of human history out to the surface to be released. The tragic experiences of our forefathers have imprinted themselves into the fabric of our DNA for millennia and they must now be shaken loose. These old energies will destroy us if they are allowed to play out. We are not fighting with instruments of individual destruction; we are holding the entire globe for ransom with just one nuclear weapon.

Certain places on Earth are the focal points of eruption, but the seething is a tension that is pushing the genetic strands of all humanity. There are no good guys and bad guys; there is only the lonely human looking for the safety of home and family.

It is difficult for us to take responsibility for the "unspeakables" of others, but until we do, we will not emerge from this shadow land of the past. There is a way to take responsibility without faltering under the burden of the past. It is to be responsible for making the changes so that we do not return to the past, to lift us so that we are freed of the illusions and transgressions that have caused such suffering.

No one on the planet is further away from you than your 44th cousin. Your genetic material is part of a global fabric and you are a conglomerate of the fibers from all the different families and races woven together within our species. Whatever your ancestors were involved in, mine were right there with them. Neither you nor I want to claim any part in these terrible things, but in our deepest recesses we are part of them. We have inherited them—and only we can heal them!

By searching out those hidden hatreds, you can dissolve them from your DNA so that your children's children will not carry the seeds of human frailty. When you come upon the historic events in the lives of your ancestors, you can discover the compassion that arises from knowing what caused the killer to be a killer, or a conqueror, or a victim.

From a spiritual perspective, you do not need to justify or defend their deeds, but rather, realize how you can change all humanity by releasing the vicious circle of repetition. By sifting through your family's history, you will be sifting through mine as well. I profoundly thank you for that.

Through the portals of global inheritance, we will emerge as a collective entity. The human genetic pool is really quite small and alterations in one group affect the larger whole. We are conglomerates. Just as all the organs come together to make one body, our human family is bound together genetically and all members contribute to the whole. Through our consciousness, we can begin to redefine what it means to us to be so alike.

Native Americans traditionally use the drum to bring the energy or consensus of the group into one heartbeat. This is precisely the necessary choice for us now on a global level. It is not that we need to be homogenized as humans, but that we can attune to each other as a species. In so doing, we bring our world into balance and our new global consciousness will open up a myriad of possibilities beyond our present scope of comprehension. What will it take for us to reach this colossal ledge of humanness? A willingness to participate in Dr. Spock's mandate—"for the good of the whole."

The change of heart that allows us to reconnect to other humans often occurs when some tragic event interrupts our monologues of judgement and we experience compassion for their suffering while at the same time the profound gratitude that we have been spared the same fate. The goodness of humanity comes into play and we want to extend ourselves to those who are in the throes of what we are so grateful to have escaped.

We are about to discover that, in fact, "what happens to you, happens to me". Any profound experience of one group or one place ripples out across the psychogenetic airwaves and triggers a corresponding response in others, even though they may be completely unaware of the connection.

Mother Earth is giving us an excellent example of how this works. The pulse of pollution is not hindered by any preemptory human notion of national boundaries. In the past we might have been simply annoyed by some pollutive act of a neighboring state or country, but now we are faced with the frightening evidence that environmental indiscretion affects us all. There are no innocent bystanders; we are all in this together.

We have discovered that a nuclear explosion in Russia or in the South Pacific islands has invasive powers extending to completely opposite sides of the world. The rain forest cut in Brazil affects weather patterns across the globe. Nature herself has tricked the humans into becoming aware of how important it is to enter into communication with each other. She is teaching us that the most important of our conversations is the conversation of our place in the planet as a whole.

The great oil spills that have poisoned our seas and the nuclear accident at Chernobyl have left no doubt about our common fate in the event of careless human actions. There has been some good that came from all this adversity. Humans have reached across the abyss of differences and helped each other. Through our cooperation, we have learned respect and caring. These seemingly isolated acts of teamwork are preparing us for the inevitable future of a global family.

Personally, I suspect Mother Nature of plotting the whole drama in order to teach us how to get along. In the end, if humans can't get the basic game plan, she will simply remove us. Time is on her side. Change is her yardstick.

As the world shrinks into a global neighborhood, we are becoming excruciatingly aware of each other's daily realities. We can hardly turn our backs on hunger, disaster, or war. It would be impossible to pretend we do not know about events in other parts of the world now that we have instant satellite reply.

My students at the Nizhoni School actually use their intuitive skills to read world events from a precognitive level. Our bodies are like Geiger counters that measure changes in earth movement, atmospheric pressure, and even human violence. The Nizhonies then tune into the feelings and respond by extending color to that place or people where the specific event is occurring.

I discovered many years ago that about 24 to 48 hours before an earthquake, my body would begin to "slosh" in a certain way that made me feel almost sick to my stomach. It was a very reliable measure of seismic activity. Many people experience various kinds of physical phenomenon in response to the Earth. It is interesting that the seat of the emotional body is the solar plexus. That is the area of your stomach and umbilicus. The solar plexus nerve ganglion interprets incoming distress signals—whether they are the instability of Earth or fear coming from another person, and directs them to the brain via the vagus nerve that solicits the reaction pattern.

Our psychogenetic pathways initiate us into realities sculpted by emotional currents. It is imperative that we learn how to use them to link us to each other and to all of nature's life forms. All humans experience the same range of emotions; the laughter, tears, passions, and peace are the currency of life's exchanges between families and communities.

The greatest gift of humanity is to experience the power of emotion: the power of the human heart. The difficulty is that our relationship to our Emotional Body is one of anxiety and longing, rather than the conscious energy of our higher emotions that would link us to states of absolute grace. Humanity is pushing up against a silent wall that hides the infinite cosmic love whispering gently to us from across the veil. We know that it is there, yet we cannot find a way of experiencing it personally. We have forgotten the oneness of our Soul and thus we translate all experience into personal ownership. We attempt to extract it from our relationships and cannot conceive of it as a force shared by all humans, but it is! This is one of our global mandates that we will answer to in times to come.

This whole planet is in a process of transmutation. We ourselves are mutants. We must recognize who we are without fear, and come to the great cosmic play with the ecstasy and pleasure of joining in with the creative force. Every genetic mutation brings about the opportunity for the kind of genetic illumination that becomes the guiding force of evolution.

There are exquisite new frequencies available to our global family. Peace is an energy that has not been present on our planet since before recorded time; perhaps only known in group to the light beings and angelic realms. At Nizhoni we say, "Peace is a choice." We are suggesting that the vibration of peace could become a crescendo that builds from individual voices to a collective harmonic orchestration. Once we choose peace, we must learn how to activate it at will and how to live it. Inner peace is the beginning point. It is not something we need wait for someone else to bring to us; it is there within each of us.

In the past we have felt that peace was a state of mind that could only be achieved through aloneness. Now is the time for us to learn that it is something we can share with others through our own peaceful essence. Peace is not really a static force; it is an essence energy that is born of our spiritual DNA. Not only can we access our inner peace, but we can entrain it within our genetic matrices so that it will be inherited by our successors.

Peace has many facets and you can perceive it in many different ways. Each time you do an exercise in consciousness to access it you may experience it differently.

Ask your body where it is holding the energy of inner peace at this moment.

Bring your consciousness into that place in your body and allow yourself to be encompassed by peace.

Imagine that the energy you experience is flooding through your entire body. Envision it being imprinted within "the mind of the cell" of the trillions of cells in your body.

In the same way that you can laser radiant white light into your DNA to alter or activate a quality, you can initialize the point of peace at its DNA site and initiate a lineage of peaceful global beings. Every time you use your consciousness to bring forth a peaceful state, it will be carried out over the ethers, across the planet and even out through our atmosphere.

All of our thoughts and actions are seeding the next ripple of humanity in the same way that the thoughts of our ancestors seeded us. Have you ever considered that the brilliance of the world's greatest beings is available to you through the collective pathways of global inheritance? The great thinkers, musicians, scientists and healers have left us the legacy of their breath and their consciousness. All we have to do is tune in to the frequencies that support those energetic qualities to begin receiving the inspiration they model for us. Every person you admire serves as a blueprint that you can adapt to your own psychogenetic design.

Contemplate your highest attributes and qualities. What is the most wonderful thing about you? Are you compassionate? Do you really love people? What would you want to extend to the world? Imagine that you are sending these energies out to seed the planet.

We are a Soul group that has chosen each other and this pivotal timeframe to bring about the changes that will secure a place in the future for us all. It is essential that we begin to think of ourselves as the family of humanity.

What will be the fate of humanity? We are creating it now.

We are the ones setting up the encoding for future generations who are coming towards us now in the same way that light is coming towards us from the cosmos at this moment. It will make it easier to feel that we can give to others when we are able to experience that we are receiving new energies ourselves from the highest source. The light awakens our consciousness, our embodiment and our Soul.

Our fate will be the result of our choices; that we have the freedom to choose is the most precious of humanity's gifts!

Chapter 11

THE TREE OF LIFE

I always think of the sea as our mother because she was the great, wet womb of our first cellular life, but she herself was born of Earth. Mother Earth birthed us all as single-celled beings close to four billion years ago. Conceived through the fusion of elements and their transformation into living structures of DNA strands, we are even now the prodigies of that one source.

The elements that merged and forged life encompassed us in a kind of vital medium that nourished the evolution of our exquisite forms. Our "tree of life" grew from the rich primordial beginnings of our planet and reached upward, responding to and altering the ever-changing texture of life on Earth.

Today, this infinite diversity still holds a genetic thread of essential components that has linked us to our common ancestors over millennia. A mere 200,000 years ago our species, *Homo sapiens sapiens* stepped forth with new awareness onto the platform of successful species.

Ours is a fantastic story that traces the tree of life from our simple beginnings as algae, through the auspicious moments of mutational changes that brought us to the present. Somewhere in the cosmic pulse, a new self-awareness emerged and we found ourselves "thinking." This marked a point of departure that we still claim separates and elevates us from all other forms of life. In a preposterous show of self-delusion, we humans have concluded that ours is the only consciousness to exist, both on planet Earth, and throughout the universe. Some people believe that God gave us this special gift and others feels that it was the result of "seeding" by more advanced beings from across the universe. Whomever the giver, we have not yet understood how our gift of consciousness invites us to participate in the evolutionary process.

Perhaps this is the moment to review that spin point of oneness we have with all of life so that we can come full circle to the attributes we have selected and possibly re-instate some that we lost along the way. Survival adaptations that served our genesis would again be of benefit to us as we review the basic aspects of living on planet Earth. By holographically threading back through molecules and genes, we can bring forth the fantastic vibrancy that links us together. Each mutation, each new species offers diverse designs that respond to the cosmic and earthly environments.

I remember listening in awe as my son Britt recounted the cosmological events that have sparked

Earth's emergence of life. He was teaching cosmology at the Nizhoni School for Global Consciousness and was demonstrating the common ancestry of all form. He described how the cyano-bacteria invented respiration two billion years ago and later the first instance of symbiotic relationship was created when those who could exploit oxygen were assimilated by those who could not.

The oxygen-wielding bacteria became the mysterious mitochondria that may prove to be the psychogenetic threshold of all species today. A billion years ago organisms discovered that they could survive by eating each other. Each event seemed to re-iterate our story. I suddenly experienced their realities as if they were my own. Indeed, they are actually a part of us all.

Although evolution is not confined to linear time, there is a great deal we could gain by re-anchoring our genetic format back to our evolutionary past. We have carefully studied the three kingdoms of life—the mineral, plant and animal—all the phylum and species, but we have only begun to dream the intricate coordinates they hold for each other. We recognize the adverse and symbiotic, the interdependent and distal relationships that crisscross between life forms. We have systematized, categorized and accounted for them all, yet we haven't really ventured into the exploration of the most simple and obvious meaning of their relativity.

Beyond the barriers of our diversity lies a common essence that has a potential for communion we have not thought to explore. Whether mineral, plant or animal in design, we are all made of carbon atoms that draw us together in a cosmic language of community, if not actual interdependence. That language is the history of our evolutionary existence from the first primordial soup of life. It holds the encoding of a trillion adaptations that marked survival for each unique life form.

Locked in our particular species reality, it has not occurred to us that we might borrow the mutations and adaptations of others to further our own survival capacities. Because we are structurally different, we have not realized that we could go beyond genetic ordering to the octaves of essence and reconfigure the sequences that apply to certain capacities.

Years ago I thought about the potential of extracting a copy of the enzymes produced by grasshoppers in the presence of radiation. I contemplated how we could then create a vaccine for ourselves and become immune to the increasing levels of radiation impinging upon our planet.

Now, I am absolutely certain that we can lift the genetic encoding directly, without cumbersome technological interference. How could it be done? Through a kind of "species speak" in which we access our genetic relatedness by virtue of the tree of life to the grasshopper people or others who have developed this radiation tolerance and introduce it back into our genetic chemistry.

It would be a biochemical language of light and electrical fluidity that is, in fact, spoken by everyone in the animal and plant kingdoms. While we think it is impossible to communicate with other species, it is only the configuration of our thought processes that holds us back. Our genetic languages are infinitely compatible!

Mother Earth is influencing our evolution by causing us to experience new and subtle secrets stored away in her infinite body. From openings within the earth and sky, we are receiving energies and messages from other sentient beings. Their more fluid relationship with the power of nature allows them to expose us to cosmic laws we do not yet understand. Yet, as we witness these, we are becoming inspired to search deeper into the mysteries of the Universe. As we do, the depth of our present despair will drop away and we will enter into a whole new cycle of partnership with nature.

The possibilities are quantum and exponential! We have always known that sentient beings could extract intelligent conversation from members of both the animal and plant kingdom. Their responses have been verified by electronic measurement that nullifies our separate worldview.

Children and psychics have perpetually reported communication with other species, animate and inanimate. I feel that they access the biochemical coordinates by way of their emotional fields, which become the backdrop for translation. When I discovered that the gene settings are biochemical, I saw the key to the profound power of emotion.

I am sure that if you have a pet, whether it is a dog, a canary or a fish, you feel that it has an uncanny way of sensing your emotional states. You have probably had at least one experience in which you felt that your pet perceived your energy and tried to comfort you. Sometimes they do funny things that appear to be deliberate attempts to cheer you up. It is almost as if you know what your animal friends are saying to you—and vice versa.

Through *psychogenetics*, we can lift our communications onto a new octave within the medium of conscious species genetics in which information of a crucial level can be exchanged. What else might we learn and then apply?

There are many magnificent life forms that have much to teach us. The whales are very inspiring to me. I wonder how they outgrew hostility and what a brain perceives that never completely sleeps. The whales evolved two million years before us and developed a brain that is essentially like our own but capable of holographic awareness. Their sonar consciousness interacts with multi-dimensional worlds. In their presence, I have felt the vastness of a consciousness not dependent on sight, but so expanded that I was carried out to the reaches of the cosmos.

Just imagine what we could do with adaptive information other remarkable species have developed. Sea anemones, for example, are said to never age. How do they do it? Viruses can alter their DNA in the face of hostile energies. Let them teach us! What do dolphins actually hear? How do dogs and other animals seem to know when an earthquake is imminent. And on and on.

I wonder at the marvelous invention of the blue-green algae, our primary ancestor, to have discovered the miracle of oxygenic photosynthesis and the utilization of gases that gave us oxygen and therefore—life. Their capacity to take in carbon dioxide and convert it to oxygen completely changed the atmosphere of our entire planet. What a fantastic feat for such a seemingly primitive organism!

Several billion years ago the blue-green algae coalesced itself into the very root of our tree of life. From its source has come every other form of life, from flowers to insects, mammals, and MAN. The blue-green algae provide us with the link to viruses, plants and animals because they are the precursors of all three. They may also be the perfect source of information about algae colonies across the universe!

We know that we are descendants of the blue-green algae but we perceive that fact more as a technicality than a relationship. In truth, we can hardly acknowledge a connection to anything but humans and we are rather poor at that. Perhaps our frail communication links stem from our habit of expediency. If we can't see what we would get from someone, we don't bother. Our attitude has left us alone and isolated from the infinite life forms and sentient beings that could enrich our world.

What if we could extract the algae's secrets! They would not resist the sharing, as we are the legacy of their "immortality". In fact, the genomes of every living form are available to us all, but we have not known how to ask the questions that would unveil the genetic encoding relative to form and function. We have reached a temporary zenith of our species and are poised for a monumental evolutionary leap into a new kind of Homo sapiens.

Other species may be the link to frequencies or dimensions that we need for our new bodies. Perhaps the grasshopper people, the lizards or countless others are doing something we are not aware of. We need to expand our creative awareness to find answers and realities that we haven't before.

We have discovered that the mitochondria of the algae communicate from one cell to another via some kind of telepathic flashes. We know that they store ATP within the cell and thus are the oxygen sparklers of life. Perhaps there is a correlation to the activities of our own mitochondria. It is not inconceivable that our mitochondria could perform the same function between us and open up an encyclopedia of survival directives that would be of great benefit to us all.

The blue-green algae developed beta and other carotene complexes to protect them from the massive radiation levels of early planetary existence. Even if we are taking blue-green algae supplements (which we should be), it behooves us to discover how to tap into their frequencies and do the same. First we must be willing to imagine it, to stretch our consciousness to conceive of the images or the holographic knowing that comes from these other life perspectives.

The whales are perhaps the easiest target for this communing venture, because they are mammals like ourselves, and their water environment is the best transmitting vehicle that exists. Our brain is 85% water and thus will allow us to tune in easily. Since they are sonar in consciousness, simply focusing on them will begin to open the "higher mind," our holographic brain coordinates which we have inherited from them. Would you like to try it?

Close your eyes. Take several deep breaths. Breathe into your brain and begin to focus on the water element of your brain, perceiving its fluidity.

When you feel that you are submerged in your brain, extend your consciousness out and connect with the brain of a whale. See what happens. (Some people begin to hear sounds. Others feel as if their heads are spinning. Others feel that they become the rhythmic pulses of sonar waves.) Stay in this energy as long as you can.

Now ask your DNA to show you exactly the spin point of sonar consciousness that you inherited from the whales.

Ask your Higher Self to laser the exact frequency of white light to activate and amplify that inheritance–now.

Feel this new energy surrounding and encompassing you.

I feel that the whale consciousness is a part of our future, not our past. Possibly, this is a species that is so beyond the doing in form that they live in absolute peace. We cannot imagine what they are creating with their sonar consciousness. It holds no physically visible images, but may be a composite world of a different kind of structure. Are they listening to the cosmos? Did they come from there? I, for one, want to know about their worlds!

Holographic consciousness is definitively an adaptive measure we need for our future. Vision is wonderful, but it leaves out simultaneous realities that we humans need to utilize—or perhaps we will develop a more multi-dimensional and transparent visual field that will transpose before us the worlds within worlds that we sense are there. The future of healing and evolutionary adaptation is consciousness!

Chapter 12

DEVIC, ANGELIC AND GALACTIC DNA

The tearing open of the veils of invisibility has presented us with the inexplicable, yet indubitable existence of infinitely more advanced life forms than ourselves. We are fascinated and frightened that there might be others who could wield more power than ourselves and we presume that they would misuse it. We fantasize them as monsters and aliens, yet in all our conjuring, it has not occurred to us that we might be related!

As we begin to sojourn out into our solar system, it is becoming obvious that whatever is out there has participated in our creation or has influenced our evolution is some mysterious way. We have always insisted that everything and everyone on Earth originated here. In fact, the cosmos is filled with energies that are on their way from one place to another. Whether they have been exploded out from a star, or have come to rest and taken root within the genetic matrix of another life form, there has been a stirring of the cosmic soup that sources the rich spectrum of living forms.

Our beloved planet Earth holds the same core carbon atoms that lace all life together within this solar system and beyond. It bears possibility then, that carbon structures from elsewhere can and *have* adapted here. Are we the result of such a transplantation? No one can say, as yet.

The line between the physical and the nonphysical has always been confusing to us. We cannot conceive of a life form outside the third dimension that could carry DNA. If we cannot see it through our microscopes, we cannot reconcile the concept of a biochemical pattern of DNA molecules that describes it, but they exist. It is probable that biochemical energies behave in different ways within astral or cosmic frequencies than they do when they are bound to the energetic laws of the third dimension.

As our common ancestor, the algae give us access to all the other kingdoms. Imagine the untold genetic possibilities that exist throughout the dimensions and the cosmos. Our human genetic matrix is intertwined with Devic, Angelic and Galactic DNA encoding, which spans the entire gamut of the astral dimension out into the cosmos. These three kingdoms have forever been integral to our evolution and we are now able to perceive them through the portholes of our higher consciousness.

All of us have some combination of Devic, Angelic or Galactic inheritance. I have illustrated some of their predominate physical indicators in my book, *Soul*

Bodies. It is amazing to look into the mirror and see ourselves in a way that includes more than just human features or interpretations of who we are.

As we come to recognize our emotional and spiritual DNA, we will find it easier to identify specific qualities inherent in each of the Devic, Angelic or Galactic kingdoms as well. Their attributes may serve as spin points of intersection that allow us special abilities or even latent aspects that may mature at some point in our life.

Without knowing how or why, we are finding ourselves surrounded by their amplifying forces. We are ready to receive them and learn what they have to teach us about our cosmic heritage and ourselves.

Can you imagine yourself having a link with any of these three groups? Many people have felt the profound relief of perceiving how this awareness explains deep feelings of longing and even familiarity with these seemingly abstract energies. Answers to strange or confusing experiences come forward when we explore our relationship to these kingdoms.

As we have now uncovered the coalescing structural matrix of DNA, we can begin to direct our consciousness to the interface between the Devic, Angelic and Galactic life forms and ourselves. Though we have no technological mechanism for tracking the convergence of our DNA, we can sense our relationship to them. In fact, we are seeded with their encoded source material. Each of them holds special gifts for us if we are willing to extend our consciousness into their frequencies.

The Devic Kingdom is composed of the Spirit of nature. Natural laws are wielded by those Devic beings whose consciousness is capable of influencing all natural phenomenon. The Devic orchestrates the weather, interspecies communication and the force of nature. The Devic Kingdom will teach us to connect again with nature spirits as we once did and from these beings we will learn how to harmonize with our planet as it, too, transitions into higher cosmic octaves.

The Angels are our cousin species who hail from the highest reaches of the astral dimension. This is why we feel so connected to them. Through our earthly Devic links, we are actually quite familiar with the astral dimension. Though it is outside of time and space, it holds the darkest possessive energies as well as the blissful, angelic heavens. We dream of both, because they are both part of our human repertoire.

From the Angelics we can learn compassion and the higher potential of relationship sourced in spiritual laws. In this next millennium, we will free the Angels to evolve out of the astral dimension and we too, will access these higher energies.

The Galactic realms are the cosmic windows that can most affect human evolution. The intense increase in contact with beings from beyond earth is necessary so that we can realize how earth is a part of an infinitely greater whole. Almost every day there is a new discovery confirming what was considered mythological just yesterday. For example, spaceship sightings in Brazil seem more plausible in light of recent findings of life elements on other planets.

Such beings have come before at auspicious moments and seeded our evolution. They are here now because we are ready for a catapultic leap. However, it is not something outside us that is changing, it is something inside us. Our DNA is calling them to us through the resonance of our like frequencies: in effect, they are inside us! We are they—in human form.

Both the need to belong and the deep longing for something beyond our own reality have been human themes throughout our history. We hunger to belong to our human family and yet we feel a strange longing for a relationship with these Galactic worlds. The truth is simply that we actually do belong to Earth and Angels and beyond!

Since approximately 1975, there has been a huge planetary awakening in which encapsulated memories encoded into blood crystals have begun to ignite. Children who embodied at that time and afterwards, have less karma and a wider window of consciousness to support their life purpose. They are already the forerunners of the new human. What these encodings imply is monumental!

We are rising out of a pit of darkness in which we saw only the shadows of our most urgent fears. Suddenly, our conscious awareness has alighted upon the possibility that we might find the solutions to our earthly problems through the interface between our limited reality and the reaches of cosmic source.

It is not that some superior group will resolve our dilemmas, but rather, as we activate our spin points that allow access into other dimensions and frequencies, we will be able to use our expanded consciousness to holographically reconnect the questions and the answers. We will "see" a new kind of truth.

Our version of sequential reality has been that

anything "alive" pertains to the interim between points of entry and exit, i.e. birth and death. It is going to be revolutionary for us to grasp that not all living forms behave in this manner. In other dimensions, the coalesced consciousness may span millennia, shape-shift, or enter into already existing bodies. It may take on a form simply as an aspect of moving back and forth across the dimensions. These beings are not actually bound in their bodies, but rather harness them for designated purposes. This is true for us as well, we just have not understood how to use it.

Many beings have the capacity to reconfigure molecules into coherent forms and then dissolve them—at will. Devas are especially adept at this, though the same is true in other kingdoms.

A tree spirit may choose the fluid consistency of its sap to conjure a form that speaks for the tree and then returns, via a "melt down" procedure, to a simple substance. In effect, it is the same process that happens to us as we melt down into body, sparking a kind of molecular tagging of cosmic consciousness that shapes itself into a human. We are actually divine essence that takes on human form in order to experience something evolutionary for the Soul.

Because we do not understand this process, we often consider these experiences to be indicative of mental disease. We label people schizophrenic who hear voices or see things that do not exist in our realm. However, it may be that the energies do actually exist, they simply arise from another dimension. It is the quality of energy that makes the difference between sickness and a valuable sensitivity. If the energies are negative, then these people are caught in a terrible psychic, emotional trap and need help. If the energies are informative and impersonal, they have a talent that needs to be trained.

The seeding of higher frequencies will change the potential of all humans. As our children begin to demonstrate abilities that we have not used, we too, will find that we are capable of things we never dreamed of. We will receive these capacities through environmental inheritance, modeled by children and others of higher consciousness. It is similar to the athletic feats of the last fifty years, where each cycle of competition produces new records, breaking those of the past.

We need not be afraid to meet our cosmic relatives. Meet them we will, for we are as crucial to their development as they are to ours. In effect, we have each inherited the other: for as surely as we carry a bit of their DNA, they must carry something of ours. Even if it has become a recessive trait, our emotional DNA is a part of their destiny.

Let us show ourselves to be capable of the adventure of change. The thrill of this next millennium will be the vast array of interdimensional and interplanetary discoveries; the challenge will be their application into our world.

Chapter 13

CONSCIOUS GENETICS

Genetics is all the rage. Suddenly a window has opened before us and we are looking out at possible futures we imagine we could create now that we are close to completing our human genome map. In the thrill of glimpsing our collective blueprint, we are already discoursing on how we could control it.

We humans never tire of the fantasy that we could somehow change what we don't like about ourselves, or that we might effortlessly install a code that would make us better than we are. Our tenacity has held together a world view that made us feel safe in the past, but in the new scheme of things, we will need something beyond our own stubborn myopic perception of reality to ensure our survival.

The new breakthroughs in genetics have given us a vantage point between the past and the future wherein we can view who we have been and who we might become, while glimpsing the signatures of who we are now. It is thrilling that a DNA sample could possibly show us the genome of a whole being, even someone dead a thousand years. But is it truly a complete signature? The genome we have discovered is only the first ripple, the outer shell of that being. It will tell the physical story, but as yet the deeper, true hologram of the being lies outside the reach of the present visible genome design.

The genes simply set a framework for possible choices and outcomes of our Soul's selection of family and environment. They express those choices, but they are not the ultimate source of them. Their patterning is in response to the psychogenetic triggers that are forever in play throughout life, not some frozen immutable form that defines our growth like a espaliered tree. We must be careful not to conclude that we are the sum total of our genes—we are not!

The danger of the gene connection is that we might see it as proof that we are not responsible for our fate and that, in fact, we are the victims of a body that was not our choice. We may see it as concrete evidence that we are the hapless products of our inheritance and thus our imperfections are insurmountable obstacles that will ultimately lead us into the same demise as our genetic relatives.

On the other hand, we may see gene manipulation as our only hope of escaping what has been dealt to us that we don't like. We may feel justified in turning over our sacred genome to those who would bow to our demands for gene replacements that would facilitate characteristics we desire or want to have in our children. Of course, we have no way of judging what else could be also inserted in the design, inadvertently or not. If we take no responsibility for attending to our genetic signatures, we allow a cloning affect on some level since we are not taking charge. In the end, we will find that neither of these postures will be the answer to our inheritance.

Gene testing might promote a fatalistic relationship with our bodies that could hold us in a pattern of fear as to how our body will ultimately bring on our demise, rather than attempting a sense of cooperation with the body. There can be no question but that the mind can control and direct the body. We must not give up the clarity of how this works in favor of a preordained doom from which we feel we cannot escape, or must trick our way out of through some technological intervention. Ultimately, we will learn that only WE can intercede, through our own consciousness.

It would be wonderful to find that a gene might ensure health, but it is much more important to discover what influences such a gene and how it interplays with the rest of the sequencing. It is ludicrous to say that repairing a gene is the answer. If the environmental heredity through the psychogenetic channels is not altered, the body will revert back after a time to the "bad" genes.

The inheritance of negative disease factors is tied to the environment of consciousness that allows our vulnerability. We inherit disease predilections from our family constellations in direct proportion to the subtle karmic situations and themes those diseases exemplify. Thus, your inherited disease is every bit as much related to the attitudes and imprints that you took on from those relatives through your psychogenetic connection, as it is to a non-functioning gene.

It makes no difference whether you were born with an improper gene formation or whether it became that way later. The source of the disruption is not in the gene itself, but the psychogenetic environment that encompasses that gene.

It is imperative that we move on from the practice of responding to the body in bits and pieces, including separating genes, as well as focusing on symptomatic vignettes. We must learn to diagnose and heal through holographic intelligence.

As I said in my book, *The Ageless Body*, "We are not just bones and blood and flesh, we are magnificent conduits of energy that make us laugh and dance and live!" Neither are our bodies such exterior stuff alone. They are light and electric-fluidity. Above all, they are the form of the formless and the expression of relativity that binds us to all humans—past and future.

Conscious mapping of our psychogenetic pathways that we inherited from our family constellations is light years ahead of what could be traced through technology. Once we are aware of these multi-dimensional and holographic aspects, we can make the alterations that would be incomprehensible on a linear level. They will bring results that we will experience as valuable and life changing for ourselves. Without the compliment of emotional and spiritual DNA attributes, we simply cannot comprehend the encoding of our genetic matrix.

It would be as foolish to not admit the evidence of emotional and spiritual DNA, even though their existence has not been scientifically proven, as it has been to blindly clutch the idea that we are alone in the universe because we haven't seen anyone else in the endless cosmos.

We often speak of inherited temperament and character, but we never attached them to a concept of emotional DNA because we had no tools to formulate its structure. *Psychogenetics* provides us with a poignant center point from which to perceive the holographic, multi-dimensional threads of genetic inheritance.

Awareness of spiritual DNA, outside the constricted confines of religious reference, will align science back to its true purpose: anwering the mystery of life and the workings of the laws of the cosmos.

In the great scheme of things, genes are only nominally more permanent than the generations of cells they create. If they are infused with purpose they will maintain—if not, they will destruct. They are guided by a force imminently more powerful than simple biochemistry. It is easier to recognize that genes can be influenced and changed when we remember that the DNA molecules that make up our genes are fluid biochemical strands. Biochemicals are definitely pliable and reactive to environmental energetics such as emotions—ours, or the residues of our great, great grandparents.

As our biochemical awareness advances, we will be able to witness genetic realities in play. The fluid chemical parings are organized or displaced by the *energy* within the medium. That energy coalesces the power of environment and thought. The dizzying array of spinning, spiraling biochemical patterns of our DNA whisper our source. It is a source that does not die with our cells, but it is a consciousness that both precedes and exceeds us.

Science knows that the atom never dies. The atom is the core of your physical essence. That means that your ancestors' consciousness is still floating around and it is certainly connected to your genetic material. It is very important to find your Higher Self so that you have a direct connection to Soul frequencies.

Soul energies are beyond the cell. The nucleus acts as a permeable membrane or threshold that constitutes a veil through which pure consciousness flows between the unmanifest and the manifest.

Who we are does not start with our genes, it starts with our Soul!

There is a purpose to the design of each body and each circumstance of life. It is the Soul who sets the stage for growth and begs life to express its potential to play out the mandate of cosmic apprenticeship.

We can only allude to that mandate through our intuitive faculties that hold the key to our questions, "Why me?" and "Who am I?" as we try to understand the laws of inheritance. Our chromosomes and genes are sensitive to our environment and our thought forms. If we wish to find the answers to these questions, we must come to know the psychogenetic history that influences the gene.

As we explore more deeply the reaches of psychogenetic inheritance, we will be able to distance ourselves from the imprints of our ancestors. Soul lessons through the body will leave disease and sorrow behind as we learn to express the miraculous potential of bodies consciously reconnected with Soul frequencies.

As a species, we were not given consciousness to simply discover our physical genetic blueprint as if that were our true story. We were given the capacity to listen to the wisdom of the body as it expresses the Soul; to discern through our awareness the solutions that will carry us forward. Our Soul offers a body that promotes the choices which will help us master the lessons we have chosen to learn from the spiritual perspective. We do not have to learn those lessons through suffering and pain. The elevation of suffering and pain as a show of spiritual worthiness is an old habit. It is time to release the concept of martyrism as a primordial throwback to a less enlightened time.

The moment that we contemplate or move inside ourselves, we have the power, the capacity, and the right to change our genetic structure so that what we pass on is the highest octave of our consciousness. No outside force can do this for us. It is the privilege of our individuation and it must be accomplished individually through the focus of each person on his or her psychogenetic blueprint.

It will take some determination to pass through this funnel of initiation. It is something that must be felt intuitively by the Self. The outside world will call for proof and refuse to entertain that no special gadgetry is necessary to change our genetic encoding. We need not wait for the future bio-technologies that can verify the existence of emotional and spiritual DNA to alter them ourselves. We are not driven by genes that are isolated in a closed circuit. Genes are shaped by consciousness. We can discover that consciousness and reinvent ourselves.

We are just emerging from an extended era of

darkness in which the human psyche was infused and deluged with terrifying admonitions coming from all authorities who wished to control the masses through the stun gun of fear. We are still afraid to choose for ourselves. Even the thought of free will immobilizes us.

Religions have been the worst offenders of this fear mongering and we must now insist that they stop the childish scare tactics and return to the divine teachings that were offered humankind from above; all of which were about the profound and loving laws of the universe. None of which were about doing what any outside authority dictated, or if not, facing the consequence of going to "hell."

We have been given magnificent templates as models for healing and experiencing ecstatic divine communion. Our spiritual DNA grants us access to these frequencies and the inheritance of such great acts performed by the Illuminated ones, so that we ourselves can follow in such blessed footsteps and do the same!

Our cellular memories can take us in many directions; directions that lead us to greatness by utilizing the foundation of the ones who seeded us, or tether us to the choices of others caught in their lonely automated bodies. Our perceptions are actually genetic habits. We can make new genetic habits that allow us to be more available to what is going on. We humans have been in a hereditary groove. We now have the choice of dissolving that groove and all the elements within it that no longer serve us.

Freed from our evolutionary lull, we can begin anew at light speed to construct a "housing" that suites our great human spirit and our Soul's wondrous humor. *Conscious genetics* is our legacy,which we will deliver to future humans by practicing it now and transforming our experiences into the constructs of our physical, emotional and spiritual DNA that they will receive as our contribution to evolution.

We have come to the edge of the world. We are standing right there on the precipice waiting for the whisper of courage to give us the strength to jump. If we align ourselves to the power of change, we will leave the limitations that beset our forefathers and soar freely into a New World. Change will become our constant and through its improvisor style, we will instigate a new pulse. It will be a pulse that rides the crest of cosmic currents flowing outward across the galaxies, taking the human heart energy with it to seed a human being we have not yet met!

About the Author

CHRIS GRISCOM

Chris Griscom is a visionary and a spiritual teacher of global stature. Her great love for the Earth and all its peoples has prompted her to travel throughout the world, reminding us all of the sacredness of life. Her teachings and exercises in consciousness have been encapsulated in twelve books, translated into thirteen languages and read by millions across the globe. As the founder of The Light Institute and The Nizhoni School for Global Consciousness, Ms. Griscom has guided people of all ages towards developing a relationship with their Higher Selves and inspired them to seek the meaning of life and to share their gifts with the world.

THE LIGHT INSTITUTE

Nestled in the magical hills of Galisteo, New Mexico, The Light Institute remains timeless. Founded in 1985 by Chris Griscom, this center for spiritual healing and multi-incarnational exploration is without equal. Individuals from around the world are attracted to the profound healing offered at The Light Institute. The Light Institute process focuses on clearing cellular memory. In a sacred and peaceful environment, clients are guided through a wonderful journey, which introduces them to their Inner Child, their Higher Self, and their own brilliance. Each session is specifically tailored to the highest growth for the participant. The Light Institute Facilitators are an international group specifically trained by Chris Griscom.

The Light Institute offers private multi-incarnational sessions and year round group intensives relating to specific themes. The Facilitators guide participants through sessions such as "Clearing the Parents," "Sexuality," "Sense of Success," and others.

NIZHONI SCHOOL FOR GLOBAL CONSCIOUSNESS

Nizhoni School for Global Consciousness offers a new form of education which allows people of all ages, from around the world, to discover their inner wisdom and bring it forth to teach, heal, and lead in the 21st century. Nizhoni was founded in 1989, by Chris Griscom, as an education of the heart, teaching holographic awareness to help us find our place in this modern world.

Nizhoni teaches that all humans are innately spiritual and that spirituality is the key to discovering our greatest gifts for ourselves and humanity. This knowing allows us to embrace all religions, all peoples thus creating new possibilities and expressions of communication and education for our communities and our planet.

Nizhoni students are taught to commune with their Divine Higher Self, thus learning to trust their own inner voice. Nizhoni is a "Soul-Centered" education; embracing the essential core of our humanness, whilst teaching how to access our highest potential and life purpose.

Nizhoni includes an international day and boarding school for primary grade and higher education students (Nizhoni Experience, The College of Divinity and The Academy of the Media). Nizhoni, exemplifies conscious living, providing a practical arena for the brilliance and extraordinary creativity that are unleashed by the Higher Self.

For more information about The Light Institute and books, tapes and videos by Chris Griscom, please contact:

The Light Institute
HC 75, Box 50
Galisteo, New Mexico 87540
USA
Phone: 505-466-1975
Fax: 505-466-7217
Websites: www.chrisgriscom.com
www.lightinstitute.com
E-mail: thelight@lightinstitute.com

For more information about The Nizhoni School for Global Consciousness, please contact:

The Nizhoni School for Global Consciousness
HC 75, Box 72
Galisteo, New Mexico 87540
USA
Phone: 505-466-4336
Fax: 505-466-7217
Website: www.nizhonischool.com
E-mail: nizhonischool@oneworldonline.net